GUIDELINES
for
Hazard Evaluation Procedures

Prepared by
Battelle Columbus Division
for
THE CENTER FOR CHEMICAL PROCESS SAFETY
of the
American Institute of Chemical Engineers

American Institute of Chemical Engineers
345 East 47th Street, New York, NY 10017

ISBN 0-8169-0399-9

It is sincerely hoped that the information presented in this document will lead to an even more impressive safety record for the entire industry; however, neither the American Institute of Chemical Engineers nor Battelle can accept any legal liability or responsibility whatsoever for the consequences of its use or misuse by anyone.

GUIDELINES FOR HAZARD EVALUATION PROCEDURES

PREFACE

The American Institute of Chemical Engineers (AIChE) has a 30-year history of involvement with process safety and loss control for chemical and petrochemical plants. Through its strong ties with process designers, builders, and operators, safety professionals, and academia, the AIChE has enhanced communication and fostered improvement in the high safety standards of the industry. Their publications and symposia have become an information resource for the chemical engineering profession on the causes of accidents and means of prevention.

Early in 1985 the AIChE established the Center for Chemical Plant Safety (CCPS) to serve as a focus for a continuing program for process safety. The first CCPS project was the preparation of guidelines for hazard evaluation procedures. The goal of this project was

> "to produce a useful and comprehensive text prepared to
> foster continued personal, professional and technical
> development of engineers in the areas of chemical plant
> safety, and to upgrade safety performance of the industry.
> It will cover the methods of identifying, assessing, and
> reducing hazards, including evaluation and selection of
> methods for particular applications. It will contain an
> executive summary, a description of the various methods,
> and useful appendices. The document will be updated
> periodically, and will serve as a basis for additional
> related topics such as risk management."

This document, Guidelines for Hazard Evaluation Procedures, is the result of the first AIChE CCPS project. It describes a selection of hazard evaluation procedures for identifying and evaluating process hazards. They include the procedures that have been developed, adopted, and used over the past 10 years by many chemical and petrochemical companies in the U.S., Canada, Europe, and elsewhere throughout the world. These were brought together by a Task Force of the AIChE CCPS to make them more widely available to the entire chemical industry and provide a firm foundation for the continuing improvement of process safety.

The procedures can be used in the design, construction, and operation of the process industries. They include procedures for identifying and analyzing hazards not easily identified from past experience. Battelle Columbus Division was chosen to review the procedures submitted, to add procedures from the more readily available literature, where appropriate, and to prepare guidelines for their use. The <u>Guidelines for Hazard Evaluation Procedures</u> describes procedures, points out where they are useful in the hazard evaluation process, and gives detailed instructions for their use.

The application of these procedures can be effective in the identification and subsequent management of process hazards, especially those with potential major consequences for the public. Actions based on the use of these procedures can lessen the probability of accidents with high consequences and reduce the consequences of the accidents that do happen. The primary emphasis is on qualitative procedures for hazard identification, although some of the procedures can also be used for quantitative hazard analysis, and that is noted in the descriptions. The guidelines are intended to apply to process hazards analysis, although some of the methods can be applied in other areas such as transportation of hazardous chemicals. The procedures can be used both for existing plants and for new plants.

Actions to reduce hazards and improve safety are no better than the extent to which hazards are recognized in the first place. No single procedure is "best" for all cases. Hazard identification requires very careful thought to analyze things as they are and to conceive of how they might be. The results of the logical thought processes of analysis and synthesis are compared with the desired end result to determine the nature and extent of the hazards. Procedures and guidelines are helpful. Clear thinking by informed and questioning minds is absolutely essential.

Chapter 1 provides descriptions of the important elements of an accident, from the initiating event through the system response to that event and on to the eventual consequences of the accident. It then addresses the elements of the hazard evaluation process itself, both the historical approach (where the objective is rigorous adherence to good practice) and the systems analysis approach (when there is insufficient experience and predictive methods are used to identify, evaluate, and control the hazards). This provides an overview of all the steps in the hazard evaluation process so that each of the individual procedures can be related to its purpose.

Chapter 2 describes a variety of hazard evaluation procedures that are being used in the chemical industry to identify equipment failures and human errors that could lead to accidents, and to minimize the consequences of accidents if failures or errors do occur. Each procedure is described in terms of its relationship to the hazard evaluation process, the period in the life of the plant or process when it is useful, the nature of the results that can be obtained, and the requirements of the procedure in skills, manpower, data, drawings, time, and cost.

Chapter 3 presents a method for selecting a particular procedure or procedures for a particular purpose, taking into consideration the above attributes of the procedures and other factors such as potential maximum consequence levels and opportunities for cost-effective risk reduction. In Chapter 4, step-by-step instructions are given for the hazard evaluation procedures described in Chapter 2.

This volume will be useful to both the experienced engineer who has done hazard evaluation, and the engineer who has neither training nor experience in this aspect of engineering. The experienced engineer will be familiar with the material in the introductory chapter, but may wish to skim the information to understand the terms as used in this document. The engineer can then proceed to review the procedures, their uses and requirements; the decision method presented to select the most appropriate procedure(s); and the instructions for those procedure(s). The inexperienced engineer will find in this document an introduction to the process of hazard evaluation, a description of procedures that are available, a basis for procedure selection, and instructions for their use. A Glossary of Terms and an extensive Bibliography are included.

In a document such as this, it is also important to state what it is not. It is not a complete management plan or program for process safety; it is only a description of hazard evaluation procedures that have been accepted and used by the process industry with detailed instructions for their use. It is not a decision process to determine what is acceptable risk; it only presents the tools for providing management with some of the information needed for decisions on process safety. It does not recommend specific procedures for specific needs, though it does provide guidelines for procedure selection. Additional factors specific to the application may also affect the choice of procedure.

Executives are fully aware of the importance of corporate policy statements that clearly define expected performance of employees regarding loss prevention, safety, health and environmental preservation. Management must be committed to and insist on safety performance being an integral part of the production process. With these systems in place, the procedures outlined in this report have been found to be helpful in manufacturing chemicals safely.

The authors appreciate the opportunity to prepare these Guidelines for the AIChE. They wish to thank the members of the AIChE CCPS Task Force for their advice and support.

Harold S. Kemp, Chairman--AIChE Vice President

Edwin J. Bassler--Stone & Webster Engineering Corporation, AIChE Director

Robert G. Chenoweth, Jr.--Union Carbide Corporation

J. Charles Forman--AIChE Executive Director and Secretary

J. Arnold Glass--AIChE, Chairman Safety and Health Division

Walter C. Kohfeldt--Exxon Chemical Company

Thomas H. Lafferre--Monsanto Company

James M. Norwood, Jr.--Air Products and Chemicals, Inc.

Gary A. Page--American Cyanamid Company

John L. Rivard--Shell Oil Company

John P. Sachs--Great Lakes Carbon Corporation, AIChE President

Anthony Santos--Factory Mutual Research Corporation

Stanley J. Schechter--Rohm and Haas Company

Robert A. Smith-- The Dow Chemical Company

Ray E. Witter--Monsanto Company.

In addition, the following members of the Task Force served on a subcommittee that interacted with Battelle in the preparation of the guidelines:

Messrs. Bassler, Kemp, Kohfeldt, Page, Schechter, and Smith.

The <u>Guidelines for Hazard Evaluation</u> was prepared by a Battelle team directed by Mr. William H. Goldthwaite, Program Manager, Mr. Fred L. Leverenz, Principal Investigator; and Dr. Paul Baybutt, Manager of the Risk, Safety, and Reliability Analysis Section. Principal authors include Mr. David P. Wagner, Mr. Robert D. Litt, and Ms. Barbara J. Bell. Other contributors include Dr. Joseph H. Oxley, Mr. William H. Mink, and Mr. John Wreathall.

The management team is grateful for the help of Ms. Kathleen H. Moore, editor, and Ms. Sue Murphy, Ms. Tina M. Payne, Ms. Angie Kennedy, Ms. Bonnie Georges, and Mr. Daryl Wolford of the report processing staff. Very special thanks to Ms. Barbara A. Foris and Ms. Kathleen A. Scoby for all those things that only the most dedicated secretaries can accomplish.

TABLE OF CONTENTS

Page

PREFACE   i

GLOSSARY OF TERMS xiii

SUMMARY xv

1. INTRODUCTION. 1-1

   1.1 The Elements of an Accident. 1-2
   1.2 Hazard Evaluation Throughout the Life of the Process 1-4
   1.3 Approaches to Hazard Evaluation and Control. 1-5

      1.3.1 Hazard Control Through Adherence to
            Good Practice 1-5
      1.3.2 Hazard Control Through Use of Predictive
            Hazard Evaluation Procedures. 1-7

2. GENERAL DESCRIPTION AND CATEGORIZATION OF HAZARD
   EVALUATION PROCEDURES 2-1

   2.1 Process/System Checklists. 2-2
   2.2 Safety Review. 2-5
   2.3 Relative Ranking--Dow and Mond Hazard Indices. 2-8
   2.4 Preliminary Hazard Analysis. 2-10
   2.5 "What If" Analysis 2-12
   2.6 Hazard and Operability (HazOp) Studies 2-14
   2.7 Failure Modes, Effects, and Criticality Analysis 2-16
   2.8 Fault Tree Analysis. 2-18
   2.9 Event Tree Analysis. 2-20
   2.10 Cause-Consequence Analysis 2-22
   2.11 Human Error Analysis 2-24

3. GUIDELINES FOR SELECTING HAZARD EVALUATION PROCEDURES 3-1

   3.1 Factors Affecting Which Procedure is Selected. 3-1
   3.2 Selection Process. 3-3

4. GUIDELINES FOR USING HAZARD EVALUATION PROCEDURES 4-1

   4.1 Process/System Checklists. 4-2
   4.2 Safety Review. 4-9
   4.3 Relative Ranking Techniques--Dow and Mond Hazard Indices . . 4-15
   4.4 Preliminary Hazard Analysis 4-20
   4.5 "What If" Analysis 4-25
   4.6 Hazard and Operability (HazOp) Studies 4-33
   4.7 Failure Modes, Effects, and Criticality Analysis 4-53
   4.8 Fault Tree Analysis. 4-61

TABLE OF CONTENTS
(Continued)

	Page

4.9 Event Tree Analysis. 4-82

4.10 Cause-Consequence Analysis 4-93

4.11 Human Error Analysis 4-97

APPENDIX A. SAMPLE CHECKLISTS. A-1

APPENDIX B. BIBLIOGRAPHY B-1

LIST OF FIGURES

Page

Figure 1-1.   Predictive Hazard Evaluation   1-9

Figure 1-2.   Steps in Predictive Hazard Evaluation.   1-13

Figure 3-1.   Matrix Relating Hazard Evaluation Procedures to
Hazard Evaluation Process Steps.   3-5

Figure 4-1.   Sample Preliminary Hazard Analysis Worksheet   4-24

Figure 4-2.   Continuous Process Example for "What If" Technique .   4-30

Figure 4-3.   Sample "What If" Worksheet for DAP Plant   4-32

Figure 4-4.   HazOp Method Flow Diagram.   4-43

Figure 4-5.   Sample of HazOp Worksheet.   4-45

Figure 4-6.   Continuous Process Example for HazOp Technique . . .   4-46

Figure 4-7.   Sample Worksheet for Knowledge-Based HazOp:
Centrifugal Pumps.   4-50

Figure 4-8.   Sample Format for an FMECA Table   4-55

Figure 4-9.   Representative Fault Tree Structure.   4-67

Figure 4-10. Sample Fault Tree   4-71

Figure 4-11. Matrix for Resolving Gates in Sample Fault Tree. . .   4-72

Figure 4-12. System Used for Fault Tree Example   4-76

Figure 4-13. Causes for TOP Event in Fault Tree Example   4-78

Figure 4-14. Causes for First Two Intermediate Events in
Fault Tree Example   4-78

Figure 4-15. Completed Fault Tree   4-80

Figure 4-16. First Step in Constructing Event Tree for
Example Cited in Text.   4-85

Figure 4-17. Representation of the First Safety Function in the
Event Tree for the Example Cited in Text   4-87

LIST OF FIGURES
(Continued)

Page

Figure 4-18.   Representation of the Second Safety Function in
the Event Tree for the Example Cited in Text   4-88

Figure 4-19.   Completed Event Tree for the Example Cited in
the Text   4-89

Figure 4-20.   Accident Sequences Derived from Completed
Event Tree   4-91

Figure 4-21.   Branch-Point Symbol Used in Cause-Consequence
Analysis   4-95

Figure 4-22.   Consequence Symbol Used in Cause-Consequence
Analysis   4-95

LIST OF TABLES

		Page
Table 1-1.	Elements of Accidents.	1-3
Table 4-1.	Typical Information Needed for "What If" Technique	4-28
Table 4-2.	"What If" Questions.	4-30
Table 4-3.	HazOp Guide Words and Meanings	4-36
Table 4-4.	Example of Criticality Ranking Definitions for FMECA.	4-57
Table 4-5.	Example of Equipment Failure Modes for FMECA	4-59
Table 4-6.	Minimal Cut Sets for the Example Fault Tree.	4-81

GLOSSARY OF TERMS

Accident, Accident Event Sequence
A specific unplanned event or sequence of events that has a specific undesirable consequence.

Audit (Process Safety Audit)
An inspection of a plant/process unit, drawings, procedures, emergency plan and/or management systems, etc., usually by an off-site team. (See "Review" for contrast.)

Consequence
The result of an accident event sequence. In this document it is the fire, explosion, release of toxic material, etc., that results from the accident but not the health effects, economic loss, etc., that is the ultimate result.

Event
An occurrence involving equipment performance or human action, or an occurrence external to the system that causes system upset. In this document an event is associated with an accident either as the cause or a contributing cause of the accident or as a response to the accident initiating event.

External Event
An occurrence external to the system/plant, such as an earthquake or flood or an interruption of facilities such as electric power or process air.

Hazard
A characteristic of the system/plant/process that represents a potential for an accident. In this document it is the combination of a (hazardous) material and an operating environment such that certain unplanned events could result in an accident.

Initiating Event
An event that will result in an accident unless systems or operations intervene to prevent or mitigate the accident.

Intermediate Event
An event of an accident event sequence that helps to propagate the accident or helps to prevent the accident or mitigate the consequences.

Mitigation System
Equipment and/or procedures designed to respond to an accident event sequence by interfering with accident propagation and/or reducing the accident consequences.

Primary Event
A basic independent event for which frequency can be obtained from experience or test.

<u>Probability</u>	An expression for the likelihood of occurrence of an event or an event sequence during an interval of time or the likelihood of the success or failure of an event on test or on demand.
<u>Review (Process Safety Review)</u>	An inspection of a plant/process unit, drawings, procedures, emergency plans and/or management systems, etc., usually by an on-site team and usually problem-solving in nature. (See "Audit" for contrast.)
<u>Risk</u>	A measure of potential economic loss or human injury in terms of the probability of the loss or injury occurring and the magnitude of the loss or injury if it occurs.
<u>Safety System</u>	Equipment and/or procedures designed to respond to an accident event sequence by preventing accident propagation and thereby prevent the accident and its consequences.
<u>"Worst Case" Consequence</u>	A conservative (high) estimate of the consequences of the most severe accident identified. For example, the assumption that the entire contents of a contained volume of toxic material is released to the most vulnerable area in such a way (all at once or continuous) as to have the maximum effect on the public or employees in that area. The contained volume could be chosen as the containers and pipes between shutoff valves or the entire process unit but probably not the entire plant.

SUMMARY

The purpose of this document is to make generally available the hazard evaluation procedures that are currently being used by many companies to reduce the risk of chemical process accidents. This work was sponsored by the American Institute of Chemical Engineers as a part of their continuing effort to improve the safety performance of the industry through education of engineers who design, start up, operate, and manage chemical and petrochemical process plants. The hazard evaluation procedures are described in sufficient detail in Chapter 2 that the most appropriate can be selected for a particular purpose with the help of the decision process described in Chapter 3. There is sufficient detail in Chapter 4 on how to use the procedures that many of them can be put into practice by persons with little experience in hazards analysis. When particular expertise is required, it is so noted.

The procedures described in this document can be used to identify accidents that occur infrequently but result in serious injury or loss when they do occur. In principle, many of the hazard evaluation procedures that are described could also be used to study more frequent, low-consequence accidents.

Hazard evaluation procedures have been developed to identify the hazards that exist, the consequences that might occur as a result of the hazards, the likelihood that events might take place that would cause an accident with such a consequence, and the likelihood that safety systems, mitigating systems and emergency alarms and evaluation plans would function properly and eliminate or reduce the consequences. This document concentrates on hazards that involve loss of containment of flammable, combustible, highly reactive, or toxic materials in amounts sufficient to endanger the health and safety of plant employees and neighboring public and/or cause serious economic loss. Many of the procedures can be used to provide either quantitative or qualitative results for decision making. This document concentrates on the use of the procedures for qualitative results.

Once the proper corporate policy and management procedures are in place, there are two approaches to hazard evaluation and control: adherence to good practice, and predictive hazard evaluation. Many standards, codes, procedures, rules, and other forms of good practice have been proven and accepted by the chemical process industry over the years. By making certain that these good practices are adhered to in design, construction, operation, and maintenance and in making changes in design, equipment, or procedures, a high level of safety can be achieved.

When a process or plant involves process or equipment designs, process materials, procedures, etc., that are different from established practice, an additional approach is necessary to identify hazards and potential accidents involving those hazards. This approach has been called predictive hazard evaluation. Procedures have been developed for predictive hazard evaluation which facilitate a thorough, systematic, element-by-element examination of a process, plant, or system to identify hazards and the ways in which equipment malfunction or human error could cause an accident. There are procedures for determining the severity of the consequences and for determining the likelihood that such accidents could occur. These procedures, by identifying the individual events in the accidents, often lead directly to measures which will affect those events and reduce the risk.

Hazard evaluation procedures can be used at any phase of process development, from laboratory process research through plant shutdown and decommissioning. For example, attention to hazard control in the research phase might lead to a substitution of materials that are less toxic or materials whose reaction rates are less sensitive to process parameters. In the manufacturing phase, hazard evaluation is vital whenever changes in equipment or procedures are considered which might introduce a new hazard or make an accident more likely.

Hazard evaluation procedures differ from each other in some ways and are similar in other ways. Attributes which are important to their selection for a particular application are:

- Purpose - e.g., hazard identification, initiating event identification
- When to Use - appropriate to use at particular phase(s) of process development
- Types of Results - e.g., lists, ranking of hazards, events
- Nature of Results - qualitative and/or quantitative

- Data Requirements - e.g., piping and instrumentation drawings, thorough understanding of process
- Staffing Requirements - skills, knowledge, number of participants
- Time and Cost Requirements.

The hazard evaluation procedures that are used most frequently by the chemical process industry for identifying deviations from good practice and to identify previously recognized hazards are Checklists and Safety Reviews. A somewhat different approach which also makes use of previous experience is the use of Dow and Mond Hazard Indices to develop a relative ranking of risks and an estimate of plant damage and economic loss based on the characteristics of the process materials and experience with plant layout and safety and mitigation systems. For predictive hazard analysis, Hazard and Operability (HazOp) Studies, Failure Modes, Effects, and Criticality Analysis (FMECA), the "What If" Method and, to a lesser extent, Fault Tree Analysis (FTA) are used routinely by many companies. Preliminary Hazard Analysis (PHA) is a procedure used by the military by that name and by chemical companies under other names. Event Tree Analysis and Cause-Consequence Analysis complement the Fault Tree Analysis, and although their concepts are implicit in the hazard analysis of many companies, they are not commonly used under that name. Human Error Analysis provides input to some of the other hazard evaluation procedures similar to input from data analysis of equipment failure rates.

These hazard evaluation procedures are described in sufficient detail that the most appropriate procedure(s) can be chosen for a particular analysis and, in most cases, can be performed by line or staff engineers. Where special training is required, it is noted in the procedure descriptions.

1. <u>INTRODUCTION</u>

The preparation of this document was sponsored by the American Institute of Chemical Engineers as a part of their continuing effort to improve the safety performance of the industry through education of engineers who design, start up, operate, and manage chemical and petrochemical process plants. Its purpose is to make generally available the hazard evaluation procedures that are currently being used by many companies to reduce the risk of chemical process accidents. These procedures can be used to identify accidents that occur infrequently but result in serious injury or loss when they do occur. In principle, many of the hazard evaluation procedures that are described could also be used to study more frequent, low-consequence accidents.

The hazard evaluation procedures are described in sufficient detail in Chapter 2 that the most appropriate can be selected for a particular purpose with the help of the decision process detailed in Chapter 3. There is sufficient detail in Chapter 4 on how to use the procedures that many of them can be put into practice by persons with little experience in hazards analysis. When particular expertise is required, it is so noted.

Effective hazard control requires a systematic, comprehensive, and precise analysis of the system and its operation. Unfortunately, there is apt to be considerable ambiguity in the definition of terms and structure of hazard control as it is described in most literature. This document strives to be consistent in the use of terms and in the systematic structure of the problem. This is not to say that there is universal agreement on these particular definitions and structures, but for the purpose of this document, agreement is less important than consistency.

1.1 The Elements of an Accident

The purpose of hazard evaluation is to identify possible accidents and estimate their frequency and consequences. For this purpose, an accident is defined as a specific unplanned sequence of events that has an undesirable consequence. The first event of the sequence is the initiating event. Conceivably the initiating event could be the only event, but usually it is not; usually there are one or more events between the initiating event and the consequence. These intermediate events are the responses of the system and its operators to the initiating event. Different responses to the same initiating event will often lead to different accident consequences. Even when the consequences are of the same type, they will usually differ in magnitude.

It is important to see the accident as a sequence of events because in theory, each individual event represents an opportunity to reduce the frequency or consequence of the accident. Table 1-1 presents examples of initiating events, intermediate events (system and operator responses) and consequences. The system and operator responses are of two types: propagating, where the accident continues to propagate through the system; and ameliorating, where special systems or special procedures come into play to reduce the level of consequences of the accident.

Also shown in Table 1-1 are examples of hazards that may be present in a chemical process plant. The term "hazard" has been used in many different ways. In this document a hazard is defined as a characteristic of the system that represents a potential for an accident with an undesirable consequence. This document concentrates on hazards that involve loss of containment of flammable, combustible, highly reactive or highly toxic materials in amounts sufficient to seriously endanger the health and safety of the plant employees and neighboring public and/or cause serious economic loss. The process of identifying and evaluating the hazards of a particular system/plant is one of identifying and evaluating the hazards inherent in the plant and all the important accident event sequences that can happen to the plant which involve those hazards.

TABLE 1-1. ELEMENTS OF ACCIDENTS

Hazards	Initiating Events/Upsets	Intermediate Events (System and Operator Responses to Upsets)		Accident Consequences
		Propagating	Ameliorative	
Significant Inventories of	Machinery and Equipment Malfunctions	Process Parameter Deviations	Safety System Responses	
a) Flammable Materials	a) Pumps, Valves	a) Pressure	a) Relief Valves	Fires
b) Combustible Materials	b) Instruments, Sensors	b) Temperature	b) Back-up Utilities	
c) Unstable Materials		c) Flow Rate	c) Back-up Components	Explosions
d) Toxic Materials		d) Concentration	d) Back-up Systems	
e) Extremely Hot or Cold Materials		e) Phase/State Change		Impacts
f) Inerting Gases (Methane Carbon Monoxide)				
Highly Reactive	Containment Failures	Containment Failures	Mitigation System Responses	Dispersion of Toxic Materials
a) Reagents	a) Pipes	a) Pipes	a) Vents	Dispersion of Highly Reactive Materials
b) Products	b) Vessels	b) Vessels	b) Dikes	
c) Intermediate Products	c) Storage Tanks	c) Storage Tanks	c) Flares	
d) By-products	d) Gaskets	d) Gaskets, Bellows, etc.	d) Sprinklers	
		e) Input/output or venting		
Reaction Rates Especially Sensitive to	Human Errors	Material Releases	Control Responses / Operator Responses	
a) Impurities	a) Operations	a) Combustibles	a) Planned	
b) Process Parameters	b) Maintenance	b) Explosive Materials	b) Ad Hoc	
	c) Testing	c) Toxic Materials		
		d) Reactive Materials		
	Loss of Utilities	Ignition/Explosion	Contingency Operations	
	a) Electricity		a) Alarms	
	b) Water	Operator Errors	b) Emergency Procedures	
	c) Air	a) Omission	c) Personnel Safety Equipment	
	d) Steam	b) Commission	d) Evacuations	
		c) Diagnosis/Decision-Making	e) Security	
	External Events	External Events	External Events	
	a) Floods		a) Early Detection	
	b) Earthquakes	a) Delayed Warning	b) Early Warning	
	c) Electrical Storms	b) Unwarned		
	d) High Winds			
	e) High Velocity Impacts			
	f) Vandalism			
	Method/Information Errors	Method/Information Failure	Information Flow	
	a) As Designed	a) Amount	a) Routing	
	b) As Communicated	b) Usefulness	b) Methods	
		c) Timeliness	c) Timing	

1.2 Hazard Evaluation Throughout the Life of the Process

It is apparent from the previous discussion of the elements of an accident that the consequences of an accident can be influenced by the process materials, the process conditions, the plant layout and equipment, the operator's procedures and training, the emergency plans, and the attention of management and staff to all of these. The other dimension is time, or phase of process/plant development. An evaluation of the hazards introduced by highly reactive, impurity-sensitive, or toxic materials should be a part of the concept development phase. Alternative materials should be considered and the penalties of using them compared to the cost of controls and safety systems that the more hazardous materials might require. Prior to completion of process/plant design and construction, a thorough search for new hazards and new accident event sequences should be made. Prior to startup, inspections of the plant and operating procedures using checklists are necessary to ensure that the plant as constructed is the same as the process/plant design that was intended and that was evaluated previously for hazards and accidents.

During the operating life of the plant, safety reviews will help ensure that the plant remains as designed and constructed through proper operation and maintenance. If changes or modifications are planned, they should be preceded by a thorough hazard evaluation of the changes and any other systems or procedures that could possibly be affected by the changes. At plant shutdown, the decommissioning and final status of the plant should also be examined for hazards.

1.3 Approaches to Hazard Evaluation and Control

The two basic approaches to hazard evaluation and control discussed in this document are: (1) adherence to good practice and (2) predictive hazard evaluation. It is important to distinguish between these approaches so that the hazard evaluation procedures that are described later can be related to one or both of these approaches.

Of course, adherence to good practice is a minimum requirement for any activity in the chemical process industry and elsewhere. This consists of observing the rules and regulations, meeting the requirements of the accepted standards, and following the practices that have been proven best from years of experience with the same processes, the same plant designs and require-ments, and the same operating and maintenance procedures. Hazard evaluation procedures such as Checklists and Safety Reviews are used to identify deviations from accepted standards and good practices.

Predictive hazard evaluation is an additional step that is needed when new and different processes, designs, equipment, or procedures are being considered. When experience is lacking, it is necessary to examine the system for new hazards, new potential accidents, and new ways in which the system may respond to accidents. Hazard evaluation procedures have been developed to provide for systematic examination of new processes and plants, and old processes and plants when modifications are planned which lie outside the experience base.

1.3.1 Hazard Control Through Adherence to Good Practice

Good Practices Evolve From Experience. As people have gained experience with the production, use, and handling of various chemical materials, they have also documented this experience in the form of standards or recommended procedures. There exist many standards for plants, companies, industries, or nations that document this experience as the standard procedure for many recurring or similar situations. These engineering standards summarize today's accepted good practice.

Many large companies have engineering design standards that specify how to design specific equipment commonly used in their businesses. These design standards vary in detail and form, but they generally conform to local, state, and national regulations as well as the standard practices of major engineering societies. These standards can be used with confidence in routine situations but may be inadequate for unusual circumstances or for new applications.

Companies that do not have a corporate engineering design manual must utilize basic theory, state and local building codes, Federal regulations, and societal codes as a basis for proper design. The most frequently encountered societal codes are:

- American Society of Mechanical Engineers (ASME) Boiler and Pressure Vessel code is almost universally accepted as the basis for construction of such equipment.
- American Petroleum Institute (API) has standards for common mechanical equipment plant layout and for the use of process instrumentation.
- Instrument Society of America (ISA) has standards for the manufacture, calibration, and application of process instruments.
- National Electrical Code (NEC) provides standards for several classifications of electrical equipment plus guidance on when to use each class.
- National Fire Protection Association (NFPA) has numerous standards directed toward establishing proper safeguards against loss of life and property by fire.

Other groups which provide useful standards or reference information include: the Chemical Manufacturers Association (CMA), Associated Factory Mutual Fire Insurance Companies, and ASTM.

Many of the standards and good practices are the result of lessons learned from analysis of accidents and near accidents. Also, experience in the industry has provided the engineer with literature on previously identified hazards associated with particular processes and plant designs.

Lessons learned have led to improvements in design, construction, operating and maintenance procedures, and management practices and controls. The conscientious use of engineering standards/good practice is a significant and necessary step toward identifying and managing the risk involved in the chemical process industry.

Methods for Identifying Deviations From Good Practice. There are two hazard evaluation procedures that will help to ensure that good practices will be followed for purposes of hazard control: Checklists and Safety Reviews. These are used to ensure that design specifications (based on standards and good practices) are met, that previously recognized hazards can be identified, and that operating and maintenance procedures conform to principles and practices that have evolved from experience. Checklists and Safety Reviews are described in Chapters 2 and 4. They must be specialized for each application, and many companies have developed their own checklists and audit procedures based on their own experiences. They are an effective way to take advantage of lessons and progress learned from experience, but they do not provide a creative search for new hazards when experience is lacking.

1.3.2 Hazard Control Through Use of
       Predictive Hazard Evaluation Procedures

Description of Predictive Hazard Evaluation. Predictive hazard evaluation procedures have been developed for analysis of processes, systems, and operations which are sufficiently different from previous experience that adhering to the good practices discussed in the previous section may not be adequate. They may also be used when evaluating a very-low-probability accident with very high consequences for which there is little or no experience. The concept addresses both the probability of an accident and the magnitude and type of the undesirable consequence of that accident. Risk is usually defined as some simple function of both the probability and consequence, but sometimes concern for either probability or consequence dominates the risk decision. Predictive procedures can provide qualitative or quantitative estimates of the probability and consequences of accidents and can often point out ways to reduce the risk, if that is needed. Figure 1-1 is often used to introduce the concept of predictive hazard evaluation or, as it is sometimes called, probabilistic risk analysis.

Predictive hazard evaluation assumes the plant will perform as designed in the absence of unintended events (component and material failures, human errors, external events, process unknowns) which affect the plant/process behavior. The objective of predictive hazard evaluation is to identify and evaluate the hazards and the unintended events which, if associated with a hazard, could cause an accident.

In a hazard evaluation, the first task is usually to identify the hazards that are an inherent feature of the process and/or plant and then focus the evaluation on events which could be associated with the hazards. Hazard identification procedures have been developed and are discussed in Chapters 2 and 4. The Dow and Mond Hazard Indices, Preliminary Hazard Analysis (PHA), the "What If" Method, Hazard and Operability Studies (HazOps), and Failure Modes, Effects and Criticality Analysis (FMECA), can be used to identify and document hazards.

Once a hazard has been identified, it is necessary to evaluate it in terms of the risk it presents to the employees, the neighboring public, and the company. However, first it is necessary to identify the initiating event, the intermediate events of the accident event sequence, and the nature of the consequence of each accident involving the hazard. Some of the same procedures that are used for hazard identification are useful here: the "What If" Method, Hazard and Operability Studies (HazOp), Failure Modes, Effects and Criticality Analysis (FMECA), and Human Error Analysis. Other procedures such as Fault Tree Analysis (FTA), Event Tree Analysis (ETA), and Cause-Consequence Analysis (CCA) can be used to identify the intermediate events. These are all described in Chapters 2 and 4.

After each accident event sequence has been identified, it is possible to estimate the risk represented by each accident. In principle, both probability and consequence should be considered, but there are occasions

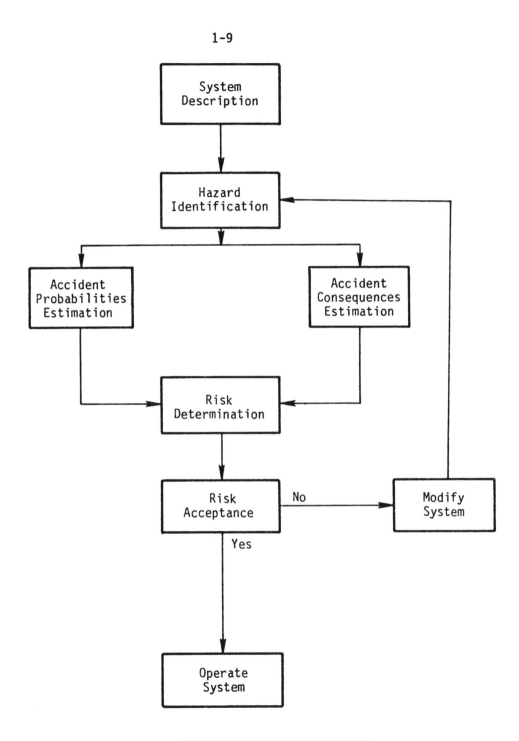

FIGURE 1-1. PREDICTIVE HAZARD EVALUATION

where if either the probability or the consequence can be shown to be sufficiently low or sufficiently high, decisions can be made on just the one factor.

The magnitudes of the consequences are estimated using engineering
analysis of the physical/chemical behavior of the pertinent process and plant
elements under upset conditions; that is, certain components have malfunctioned
and/or certain process deviations from the normal are assumed to have taken
place. The engineering analysis can be as approximate or as precise as the
occasion demands. More often than not, very approximate conservative analysis
and good engineering judgment will be sufficient for the decisions which have
to be made; however, if more precise analysis is warranted, the basis will
have been established.

The engineering analysis of plant behavior under upset conditions
provides the basis for estimating the extent of propagation of the accident
within the plant and provides the amount of release of flammable, combustible,
highly reactive, or toxic materials to the area surrounding the plant. The
Dow and Mond Fire and Explosive Hazard Indices are useful guidelines for plant
design to keep accident propagation within the plant at an acceptable level.
Chapters 2 and 4 describe these procedures. Fire Analysis and Gas Dispersion
Analysis are also useful in estimating the magnitude of certain accident consequences but are beyond the scope of this document.

The probabilities of the accident event sequences can be estimated
easily if it is obvious that the probability is dominated by a single event
and frequency data are available for that event. If that is not the case, the
pertinent systems of the plant can be modeled with logic diagrams, and the
probability of each accident can be determined from the combined probabilities
of the individual events in the event sequence. Event Tree Analysis, described
in Chapters 2 and 4, can be used for this. If necessary, the probability of
the individual events can be estimated from their basic causes with Fault Tree
Analysis. Cause-Consequence Analysis is also useful here. FTA and Cause-
Consequence Analysis are also described in Chapters 2 and 4.

Using the Hazard Evaluation Procedures to Reduce Risk. The earlier
discussion of the Elements of Accidents called attention to the sequence of

events in an accident and noted that in theory each was a point of opportunity for risk reduction. Subsequent discussions have pointed out that risk is a function of the probabilities and consequences of the accidents involving hazards and that reducing either the probability or the consequences of a potential accident will reduce its risk. Reducing the hazard, for example by reducing the inventory of hazardous material that could be involved in an accident, is a candidate method for reducing the consequence and thereby the risk.

Table 1-1 presented several types of hazards, initiating events, and system responses, each of which may be a target for reducing risk. In some cases, a reduction in frequency of hardware malfunction or human error by better components, better maintenance, redundant components, better training, better instrument displays, etc. is the avenue to reduce risk. In other cases, safety systems will stop the propagation of an accident, but only if they function properly--which addresses the probability factor again. Mitigation systems that do not affect the probability of the event sequence but do mitigate the consequences may be a cost-effective way of reducing risk.

As the hazard evaluation process proceeds, it may become obvious that the installation of a particular safety system or a particular modification to the system will achieve the desired level of risk, and the hazard evaluation could be terminated at that point. However, there are dangers in premature termination of the process. Certainly the hazard identification step should be completed, since reducing the risk of the first hazard identified will not necessarily reduce the risks of remaining hazards if there are others. Also, a "worst case" estimate of the consequences of accidents involving the hazards that have been identified should be made, with and without any changes made to reduce the risks of any of them. Finally, a change to reduce the risk of one hazard may actually increase the risk from another hazard or introduce a new hazard, so interactions of this type must be considered before the hazard evaluation is terminated.

Steps in Hazard Evaluation. Just as it is important to view an accident as a sequence of events, each of which may offer an opportunity for risk reduction, predictive hazard evaluation can be viewed as a series of steps toward predicting the risk from a plant or process. Not all steps need to be taken in every hazard evaluation, nor are the steps always taken in the same order. However, a natural order of steps is presented here and desirable variations from this order will be apparent as specific situations are encountered.

The order of steps in a typical hazard evaluation is as follows:

1. Identify the hazards inherent in the process/plant.
2. Estimate the consequences that could result from the hazards identified in (1) above.
3. Identify opportunities to reduce the consequences in (2) above.
4. Identify initiating events of accidents that could lead to the consequences estimated in (2) above.
5. Estimate the probabilities of the initiating events.
6. Identify opportunities to reduce the probabilities of the initiating events.
7. Identify the event sequences of accidents (system responses) that could lead to the consequences estimated in (2) above.
8. Estimate the probabilities and consequences of the accident event sequences identified in (7) above.
9. Identify opportunities to reduce the probabilities and/or the consequences of the accident event sequences identified in (7) above.
10. If necessary, do quantitative hazard evaluations to reduce the uncertainty in the estimates of probabilities and consequences and identify optimum investments for obtaining a level of risk deemed acceptable.

Figure 1-2 shows the relationship between these steps and the actions which can be taken at different points in the process. After each step, the results of the process will assist in the determination of whether the risk is acceptable, whether cost-effective modifications can be made, or whether the hazard evaluation should continue. It is important that the estimates of accident consequences and event probabilities be as conservative

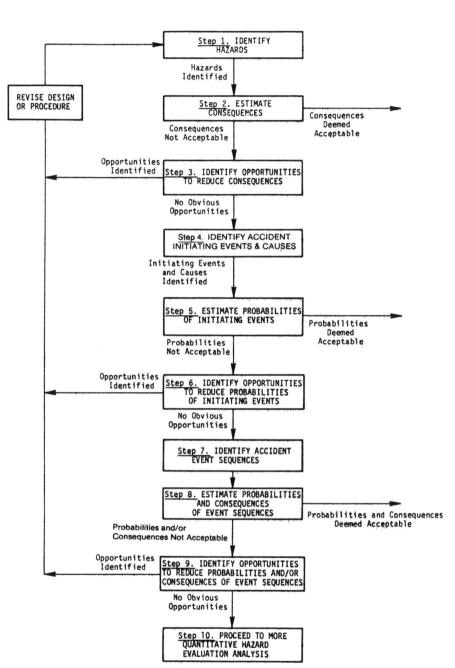

FIGURE 1-2. STEPS IN PREDICTIVE HAZARD EVALUATION

as the available data, procedures, and circumstances demand. Often "worst case" estimates are all that are justified but are still sufficient for decision making.

Hazard Evaluation Results and Follow-Up. In general, the hazard evaluation procedures described in this document will provide one or more of the following types of information:

- Identification and description of hazards and accident events which could lead to undesirable consequences.
- A qualitative estimate of the likelihood and consequence of each accident event sequence.
- A relative ranking of the risk of each hazard and accident event sequence.
- Some suggested approaches to risk reduction.

These results will be provided to plant management, engineering, or research, as appropriate to the occasion.

Follow-up is the responsibility of the appropriate management and not the hazard evaluation team. However, it is not unusual for members of the team to be consulted and even to participate in the development and selection of risk reduction actions.

Actions to reduce risk are generally of five kinds:

- A change in the physical design or control system.
- A change in operating method.
- A change in process (pressure, temperature).
- A change in the process materials.
- A change in the test and inspection/calibration of key safety items.

It is important to consider a wide range of possible actions and not to expect that every hazard can and should be contained by an alteration in physical design. When choosing among a number of possible actions, it may be useful to put them into three categories:

- Those actions which eliminate the hazard.
- Those actions which eliminate or reduce the consequences.
- Those actions which reduce the likelihood to an acceptable level.

In general, it is better and more effective to eliminate the hazard, and if the study is carried out at the design stage, this can often be done without undue expenditure. If there is no reasonable prospect of eliminating the hazard, it will be necessary to consider what can be done to reduce the likelihood of an accident or to protect people and plant if the accident takes place.

To illustrate the kind of reasoning which may be applied, consider a reaction vessel where, in a HazOp session, it was discovered that if a certain impurity were introduced with one of the raw materials, there would be a sudden evolution of gas and an increase in pressure in the vessel.

Let us assume that the hazard might be contained by taking one of three actions:

(1) Eliminating the possibility of gas evolution by changing the raw material responsible for the problem.
(2) Eliminating the possibility of gas evolution by altering one of the process conditions.
(3) Fitting an appropriate pressure relief valve and vent system to protect the plant.

Solution (1) will be 100 percent effective but may not be practical. Solution (2) has to be considered with care, because its adequacy will depend on the reliability of the control system which governs the process condition. Solution (3) is a common solution, but care must be taken that both the valve and vent system are designed to cope with the expected gas evolution.

It cannot be overemphasized that care must be taken to avoid introducing a new hazard, increasing the likelihood or consequences of a different accident, and so on. The safest course is to have the original hazard evaluation team review the new action in the light of their understanding of the other hazards present and the process and plant behavior.

2. GENERAL DESCRIPTION AND CATEGORIZATION OF HAZARD EVALUATION PROCEDURES

In this chapter, the following hazard evaluation procedures are described so that one or more may be selected for further study in Chapter 4.

Procedure	Page
Process/System Checklists	2-2
Safety Review	2-5
Relative Ranking--Dow and Mond Hazard Indices	2-8
Preliminary Hazard Analysis	2-10
"What If" Analysis	2-12
Hazard and Operability Studies	2-14
Failure Modes, Effects, and Criticality Analysis	2-16
Fault Tree Analysis	2-18
Event Tree Analysis	2-20
Cause-Consequence Analysis	2-22
Human Error Analysis	2-24

Each description addresses the purpose of the procedure, when it is used, the type and nature of its results, and its requirements in terms of data, staffing, time, and cost.

2.1 Process/System Checklists

General Description

Checklists are frequently used to indicate compliance with standard procedures. A checklist is easy to use and can be applied to each stage of a project or plant development. A checklist is a convenient means of communicating the minimal acceptable level of hazard evaluation that is required for any job, regardless of scope. It is particularly useful for an inexperienced engineer to work through the various requirements in the checklist to reach a satisfactory conclusion. It also provides a common basis for management review of the individual engineer's work.

A checklist is intended to provide direction for standard evaluation of chemical plant hazards. It can be as detailed as necessary to satisfy the specific situation, but it should be applied conscientiously in order to identify problems that require further attention and to ensure that standard procedures are being followed. Checklists are limited to the experience base of the checklist author(s). They should be audited and updated regularly.

Many organizations use standard checklists for controlling the development of a project from initial design through plant shutdown. The checklist is frequently a form for approval by various staff and management functions before a project can move from one stage to the next. In this way, it serves both as a means of communication and as a form of control.

Attributes

Purpose: To identify common hazards and ensure compliance with standard procedures.

When to Use:
a. Design: Checklists can be used at all stages of design for a quick and simple identification of the hazards involved and the appropriate means of dealing with the hazard.

b. Construction: Selected criteria should be checked during construction to ensure quality workmanship and compliance with the design specifications.

c. Startup: A rigorous checklist is often the most appropriate means of progressing through the startup period.

d. Operation: Periodic use of a checklist during plant operation is a good means of evaluating compliance with standard procedures.

e. Shutdown: A checklist prepared specifically for this often-neglected phase of any project or plant is useful to minimize any latent hazards that exist in idle or unused equipment.

Type of Results: Generally the checklist identifies common hazards and leads to compliance with standard procedures. It can also high-light a lack of basic information or a situation that requires a detailed evaluation.

Nature of Results: Qualitative results from using a checklist vary with the specific situation, but generally they lead to a "yes-or-no" decision about compliance with standard procedures.

Data Requirements:
a. Checklist prepared from prior experience
b. Standard Procedures Manual
c. Knowledge of system or plant.

Staffing Requirements: Generally, one or more experienced persons must prepare the checklist. Any engineer should be able to use the checklist, with necessary guidance based on individual experience. An experienced manager or staff engineer should then review the checklist results and direct the next action.

<u>Time and Cost Requirements</u>: A checklist is easy to use and can provide results relatively quickly. The extent of the checklist can vary, but it is generally one of the quickest and least expensive methods of risk evaluation. It is highly cost effective for common hazards.

2.2 Safety Review

General Description

A walk-through on-site inspection can vary from an informal routine function that is principally visual, with emphasis on housekeeping, to a formal week-long examination by a team with appropriate backgrounds and responsibilities. The emphasis in this section is on the latter and it is sometimes referred to as a "Safety Review". It may also be referred to as a Process Safety Review, a Loss Prevention Review, or a Process Review. Such a program is intended to identify plant conditions or operating procedures that could lead to an accident and significant losses in life or property.

While this technique is most commonly applied to operating process plants, it is also applicable to pilot plants, laboratories, storage facilities, or support functions. The comprehensive Safety Review is intended to complement other safety efforts and routine visual inspections. The Safety Review should be treated as a cooperative effort to improve the overall safety and performance of the plant rather than as a dreaded interference with normal operations. Cooperation is essential. People are likely to become defensive unless considerable effort is made to present the review as a benefit to each participant.

The review includes interviews with many people in the plant: operators, maintenance staff, engineers, management, safety staff, and others, depending upon the plant organization. Having the support and involvement of all these groups provides a thorough examination from many perspectives.

The review looks for major risk situations. General housekeeping and personnel attitude are not the objectives, although they can be significant indicators of where to look for real problems or places where real improvements are needed. Various hazard evaluation techniques, such as checklists, "what-if" questions, and raw material evaluations, can be used during the review.

At the end of the Safety Review, recommendations are made for specific actions that are needed, with justification; recommended responsibilities; and completion dates. A follow-up evaluation or re-inspection should be planned to verify the acceptability of the corrective action.

Attributes

Purpose: Safety Reviews are a tool for ensuring that the plant and operating and maintenance procedures match the design intent and standards. The procedure keeps operating personnel alert to the process hazards; reviews operating procedures for necessary revisions; seeks to identify equipment or process changes that could have introduced new hazards; initiates application of new technology to existing hazards; and reviews adequacy of maintenance safety inspections.

When to Use: Safety Reviews are usually conducted on a regularly scheduled basis. Special-emphasis reviews or follow-up/resurvey inspections can be scheduled intermittently. Some companies review "high-risk" plants as often as every 2 or 3 years. The intervals may be as long as 5 to 10 years for other companies or "low-risk" plants.

Type of Results: The inspection team's report includes deviations from designed and planned procedures and notification of new safety items discovered. Responsibility for implementing the corrective action remains with the plant management.

Nature of Results: Qualitative.

Data Requirements: For a complete review, the team will need access to applicable codes and standards, detailed plant descriptions such as piping and instrumentation drawings and flowcharts; plant procedures for start-up, shutdown, normal operation, and emergencies; personnel injury reports; hazardous incidents reports; maintenance records such as critical instrument checks, pressure relief valve tests, pressure vessel inspections; and process material character-istics (i.e., toxicity and reactivity information).

Staffing Requirements. Staff assigned to Safety Review inspections need to be very familiar with safety standards and procedures. Special technical skills are helpful for evaluating instrumentation, electrical systems, pressure vessels, process materials and chemistry, and other special-emphasis topics.

Time and Cost Requirements: A complete survey will normally require a team of 2-5 people for at least a week. Shorter inspections do not allow for thorough examinations of all equipment or procedures.

2.3 Relative Ranking--Dow and Mond Hazard Indices

General Description

The Dow and Mond Indices provide a direct and easy method for providing a relative ranking of the risks in a chemical process plant. The methods assign penalties and credits based on plant features. Penalties are assigned to process materials and conditions that can contribute to an accident. Credits are assigned to plant safety features that can mitigate the effects of an accident. These penalties and credits are combined to derive an index that is a relative ranking of the plant risk. Estimates of consequences in terms of cost and outage time can also be included in the evaluations.

Attributes

Purpose. Provide a relative ranking of plant process units based on degree of risk.

When to Use:
a. Design: Can be used on a plant design to identify vulnerable areas and specify plant protective features.
b. Operation: Can provide relative information on operating plant hazards and where safety upgrades might be beneficial.

Types of Results. Relative ranking of plant process units based on degree of risk.

Nature of Results. Relative quantitative ranking plus qualitative information on equipment exposed to possible damage through accident propagation.

Data Requirements.

a. Accurate plot plan of the plant

b. Thorough understanding of the process flow and the process conditions

c. Other requirements:

Fire and Explosion Index Form

Risk analysis form

Re-cap worksheet

Calculator and drawing compass

Fire and Emergency Index Guide

Cost data for plant process equipment.

Staffing Requirements. These methods require a chemical engineer or industrial chemist familiar with the chemistry and process unit layout. In addition, support from the company's business office will be required if the evaluation includes equipment replacement and business interruption costs. Each unit evaluation can be carried out by a single analyst who has knowledge of the process.

Time and Cost Requirements. Time and cost requirements will vary depending on the number of units in the evaluation. A qualified analyst should accomplish two to three average process unit evaluations per week, depending on the availability of the required financial information.

2.4 Preliminary Hazard Analysis

General Description

A Preliminary Hazard Analysis (PHA), as described herein, is an analysis which is part of the U.S. Military Standard System Safety Program Requirements. The main purpose of this analysis is to recognize hazards early, thus saving time and cost which could result from major plant redesigns if hazards are discovered at a later stage. Many chemical companies use a similar procedure under a different name. It is generally applied during the concept or early development phase of a process plant and can be very useful in site selection.

PHA is a precursor to further hazard analyses. It is included in this document to provide a cost effective, early-in-plant-life method for hazard identification. Indeed, the PHA is really intended for use only in the preliminary phase of plant development for cases where past experience provides little or no insight into any potential safety problems, e.g., a new plant with a new process.

The PHA focuses on the hazardous materials and major plant elements since few details on the plant design are available, and there is likely not to be any information available on procedures. The PHA is sometimes considered to be a review of where energy can be released in an uncontrolled manner. The PHA consists of formulating a list of the hazards related to:

- Raw materials, intermediate and final products, and their reactivity
- Plant equipment
- Interface among system components
- Operating environment
- Operations (test, maintenance, etc.)
- Facility
- Safety equipment.

The results include recommendations to reduce or eliminate hazards in the subsequent plant design phase.

Attributes

Purpose: Early identification of hazards to provide designers with guidance in final plant design stage.

When to Use: The PHA is used in the early design phase, when only the basic plant elements and materials are defined.

Type of Results: A list of hazards related to available design details, with recommendations to designers to aid hazard reduction during final design.

Nature of Results: Qualitative listing, with no numerical estimation or prioritization.

Data Requirements: Available plant design criteria, equipment specifications, material specifications, and other like material.

Staffing Requirements: A PHA can be accomplished by one or two engineers with a safety background. Less-experienced staff can perform a PHA, but it may not be as complete as desired.

Time and Cost Requirements: Because of its nature, experienced safety staff can accomplish a PHA with an effort which is small compared to the effort needed for other hazard evaluation procedures.

2.5 "What If" Analysis

General Description

The "What If" procedure is not as structured as some others, such as Hazard and Operability (HazOp) study and Failure Modes, Effects and Criticality (FMECA). Instead, it requires the user to adapt the basic concept to the specific application. Very little information has been published on the "What If" method or its application. However, it is frequently referred to within the industry and enjoys a good reputation among those skilled in its use.

The purpose of a "What If" analysis is to consider carefully the result of unexpected events that would produce an adverse consequence. The method involves examination of possible deviations from the design, construction, modification, or operating intent. It requires a basic understanding of what is intended and the ability to mentally combine or synthesize possible deviations from design intent that would cause an undesired result. This is a powerful procedure if the staff is experienced; otherwise, the results are likely to be incomplete.

The "What If" concept uses questions which begin "What If...". For example:

- "What if" the wrong material is delivered?
- "What if" Pump A stops running during startup?
- "What if" the operator opens valve B instead of A?

The questions are divided into specific areas of investigation (usually related to consequences of concern), such as electrical safety, fire protection, or personnel safety. Each area is addressed by a team of two or three experts. The questions are formulated based on previous experience and applied to existing drawings and charts; for an operating plant, the investigation may include questions for the plant staff. (There is no specific pattern or order to these questions, unless the application provides a logical pattern). The questions can address any variation related to the plant, not just component failure or process variation.

Attributes

Purpose: Identify possible accident event sequences and thus identify the hazards, consequences, and perhaps potential methods for
risk reduction.

When to Use: The "What If" method can be used for existing plants,
during the process development stage, or at pre-startup stage. A
very common usage is to examine proposed changes to an existing
plant.

Type of Results: Tabular listing of potential accident scenarios,
their consequences, and possible risk reduction methods.

Nature of Results: Qualitative listing, with no ranking or
quantitative implication.

Data Requirements: Detailed documentation of the plant, the
process, the operating procedures, and possibly interviews with
plant operating personnel.

Staffing Requirements: For each investigation area, two or three
experts should be assigned.

Time and Cost Requirements: Time and cost are proportional to the
plant size and number of investigation areas to be addressed. For
an organization that has not done a "What If" before, additional
time will be needed to formulate questions and accumulate relevant
information. Once an organization has the "What If" in place, it
can become a time- and cost-efficient means for plant review.

2.6 Hazard and Operability (HazOp) Studies

General Description

The HazOp study was developed to identify hazards in a process plant and to identify operability problems which, though not hazardous, could compromise the plant's ability to achieve design productivity. Thus, a HazOp goes beyond hazard identification. Although originally developed to anticipate hazards and operability problems for new and/or novel technology where past experience is limited, it has been found to be very effective for use at any stage in a plant's life from final design on. In addition, one variation of the HazOp has been developed specifically to address preliminary design.

The approach taken is to form a multidisciplinary team that works together (brainstorms) to identify hazards and operability problems by searching for deviations from design intents. An experienced team leader systematically guides the team through the plant design using a fixed set of words, called "guide words", or uses checklists or knowledge. These guide words are applied at specific points or "study nodes" in the plant design to identify potential deviations of the plant process parameters at those nodes. The nodes are usually specified by the team leader before the meetings. For example, the guide word "no" combined with the process parameter "flow" results in the deviation "no flow". The team then agrees on possible causes of the deviations (e.g., operator error shuts off pump) and the consequences (e.g., product contamination). If the causes and consequences are realistic and significant, they are recorded for follow-up action, which takes place outside of the study. In some cases, the team identifies a deviation with a realistic cause but unknown consequence (e.g., unknown reaction product) and recommends follow-on studies to determine the possible consequences.

Attributes

Purpose: Identification of hazard and operability problems.

<u>When to Use</u>: Optimal from a cost viewpoint when applied to new
plants at the point where the design is nearly firm and documented
or to existing plants where a major redesign is planned. It can
also be used for existing facilities.

<u>Type of Results</u>: The results are the team findings, which include:
identification of hazards and operating problems; recommended
changes in design, procedures, etc., to improve safety; and
recommendations for follow-on studies where no conclusion was
possible due to lack of information.

<u>Nature of Results</u>: Qualitative.

<u>Data Requirements</u>: The HazOp requires detailed plant descriptions,
such as drawings, procedures, and flow charts. A HazOp also
requires considerable knowledge of the process, instrumentation, and
operation, and this information is usually provided by team members
who are experts in these areas.

<u>Staff Requirements</u>: The HazOp team is ideally made up of 5 to
7 professionals, with support for recording and reporting. For a
small plant, a team as small as two or three could be effective.

<u>Time and Cost</u>: The time and cost of a HazOp are directly related to
the size and complexity of the plant being analyzed. In general,
the team must spend about three hours for each major hardware item.
Where the system analyzed is similar to one investigated previously,
the time is usually small. Additional time must be allowed for
planning, team coordination, and documentation. This additional
time can be as much as two to three times the team effort as
estimated above.

2.7 Failure Modes, Effects, and Criticality Analysis

General Description

Failure Modes, Effects, and Criticality Analysis (FMECA) is a tabulation of the system/plant equipment, their failure modes, each failure mode's effect on the system/plant, and a criticality ranking for each failure mode. The failure mode is a description of how equipment fails (open, closed, on, off, leaks, etc.). The effect of the failure mode is the system response or accident resulting from the equipment failure. FMECA identifies single failure modes that either directly result in or contribute significantly to an important accident. Human/operator errors are generally not examined in a FMECA; however, the effects of a misoperation are usually described by an equipment failure mode. FMECA is not efficient for identifying combinations of equipment failures that lead to accidents. The FMECA can be performed by two analysts or a multidisciplinary team of professionals.

A Failure Modes and Effects Analysis (FMEA) is equivalent to a FMECA without a criticality ranking.

Attributes

Purpose: Identify equipment/system failure modes and each failure mode's potential effect(s) on the system/plant.

When to Use:
a. Design: FMECA can be used to identify additional protective features that can be readily incorporated into the design.
b. Construction: FMECA can be used to evaluate equipment changes resulting from field modifications.
c. Operation: FMECA can be used to evaluate an existing facility and identify existing single failures that represent potential accidents, as well as to supplement more detailed hazard assessments such as HazOp or Fault Tree Analysis.

Type of Results: Systematic reference listing of system/plant equipment, failure modes, and their effects. Easily updated for design changes or system/plant modifications.

Nature of Results: Qualitative. Includes worst-case estimate of consequence resulting from single failures. Contains a relative ranking of the equipment failures based on estimates of failure probability and/or hazard severity.

Data Requirements:

a. System/plant equipment list
b. Knowledge of equipment function
c. Knowledge of system/plant function.

Staffing Requirements: Staffing requirements will vary with the number and size of systems in the FMECA and with the time constraints. For an average system evaluation, ideally two analysts should participate to provide a check for each analyst's assessments. All analysts involved in the FMECA should be familiar with the equipment functions and failure modes and with how the failures might propagate to other portions of the system/process.

Time and Cost Requirements: Time and cost of the FMECA is proportional to the size and number of systems analyzed in the FMECA. On the average, an hour is sufficient for two to four evaluations per analyst. In systems with similar equipment performing similar functions, the time requirements are reduced significantly because of the repetitive nature of the evaluations.

2.8 Fault Tree Analysis

General Description

Fault Tree Analysis (FTA) is a deductive technique that focuses on one particular accident event and provides a method for determining causes of that accident event. The fault tree itself is a graphic model that displays the various combinations of equipment faults and failures that can result in the accident event. The solution of the fault tree is a list of the sets of equipment failures that are sufficient to result in the accident event of interest. FTA can include contributing human/operator errors as well as equipment failures.

The strength of FTA as a qualitative tool is its ability to break down an accident into basic equipment failures and human errors. This allows the safety analyst to focus preventive measures on these basic causes to reduce the probability of an accident.

Attributes

Purpose: Identify combinations of equipment failures and human errors that can result in an accident event.

When to Use:

a. Design: FTA can be used in the design phase of the plant to uncover hidden failure modes that result from combinations of equipment failures.

b. Operation: FTA including operator and procedure characteristics can be used to study an operating plant to identify potential combinations of failures for specific accidents.

Type of Results: A listing of sets of equipment and/or operator failures that can result in a specific accident. These sets can be qualitatively ranked by importance.

Nature of Results: Qualitative, with quantitative potential. The fault tree can be evaluated quantitatively when probabilistic data are available.

Data Requirements:

a. A complete understanding of how the plant/system functions
b. Knowledge of the plant/system equipment failure modes and their effects on the plant/system. This information could be obtained from an FMEA or FMECA study.

Staffing Requirements: One analyst should be responsible for a single fault tree, with frequent consultation with the engineers, operators, and other personnel who have experience with the systems/equipment that are included in the analysis. The single analyst/single fault tree approach promotes continuity within the fault tree, but the analyst must have access to the information needed to define faults and failures that contribute to the accident event. A team approach is desirable if multiple fault trees are needed, with each team member concentrating on one individual fault tree. Interactions between team members and other experienced personnel are necessary for completeness in the analysis process.

Time and Cost Requirements: Time and cost requirements for FTA are highly dependent on the complexity of the systems involved in the accident event and the level of resolution of the analysis. Modeling a small process unit could require a day or less with an experienced team. Large problems, with many potential accident events and complex systems, could require several weeks even with an experienced analysis team.

2.9 Event Tree Analysis

General Description

Event tree analysis is a technique for evaluating potential accident outcomes resulting from a specific equipment system failure or human error known as an initiating event. Event tree analysis considers operator response or safety system response to the initiating event in determining the potential accident outcomes. The results of the event tree analysis are accident sequences; that is, a chronological set of failures or errors that define an accident. These results describe the possible accident outcomes in terms of the sequence of events (successes or failures of safety functions) that follow an initiating event. Event tree analysis is well suited for systems that have safety systems or emergency procedures in place to respond to specific initiating events.

Attributes

Purpose: Identify the sequences of events, following an initiating event, that result in accidents.

When to Use:

a.   Design: Event tree analysis can be used in the design phase to assess potential accidents resulting from postulated initiating events. The results can be useful in specifying safety features to be incorporated into the plant design.

b.   Operation: Event tree analysis can be used on an operating facility to assess the adequacy of existing safety features or to examine the potential outcomes of equipment failures.

Type of Results: Provides the event sequences that result in accidents following the occurrence of an initiating event.

Nature of Results: Qualitative, with quantitative potential. The expected probability of the sequences can be quantified if the event probabilities are known.

Data Requirements:

a. Knowledge of initiating events; that is, equipment failures or system upsets that can potentially cause an accident.
b. Knowledge of safety system function or emergency procedures that potentially mitigate the effects of an initiating event.

Staffing Requirements: An Event Tree Analysis can be performed by a single analyst, but normally a team of 2 to 4 people is preferred. The team approach promotes "brainstorming" that results in a well defined event tree structure. The team should include at least one member with knowledge of event tree analysis, with the remaining members having experience in the operations of the systems and knowledge of the chemical processes included in the analysis.

Time and Cost Requirements: Time and cost requirements for an event tree analysis are highly dependent on the number and complexity of initiating events and safety functions included in the analysis. Three to six days should allow the team to evaluate several initiating events for a small process unit. Large or complex process units could require two to four weeks to evaluate multiple initiating events and the appropriate safety function responses.

2.10 Cause-Consequence Analysis

General Description

Cause-consequence analysis is a blend of fault tree and event tree analysis (discussed in the preceding sections) for evaluating potential accidents. A major strength of cause-consequence analysis is its use as a communication tool: the cause-consequence diagram displays the interrelationships between the accident outcomes (consequences) and their basic causes. The method can be used to quantify the expected frequency of occurrence of the consequences if the appropriate data are available.

Attributes

Purpose: Identify potential accident consequences and the basic causes of these accidents.

When to Use:

a.   Design: Cause-consequence analysis can be used in the design phase to assess potential accidents and identify their basic causes.
b.   Operation: Cause-consequence analysis can be used in an operating facility to evaluate potential accidents.

Type of Results: Potential accident consequences, related to their basic causes. Probabilities of each type of accident can be developed if quantification is desired.

Nature of Results: Qualitative, with quantitative potential.

Data Requirements:

a. Knowledge of component failures or process upsets that could
    cause accidents.
b. Knowledge of safety systems or emergency procedures that can
    influence the outcome of an accident.

Staffing Requirements: Cause-consequence analysis is best performed
by a small team (2 to 4 people) with a variety of experience. One
team member should be experienced in cause-consequence analysis (or
fault tree and event tree analysis), with the remaining members
having experience in the operations and interactions of the systems
included in the analysis.

Time and Cost Requirements: Time and cost requirements for cause-
consequence analysis are highly dependent on the number, complexity,
and level of resolution of the events included in the analysis.
Scoping-type analyses for several initiating events can usually be
accomplished in a week or less. Detailed cause-consequence analyses
may require two to six weeks, depending on the complexity of the
supporting fault tree analyses.

2.11 Human Error Analysis

General Description

Human Error Analysis is a systematic evaluation of the factors that influence the performance of human operators, maintenance staff, technicians, and other personnel in the plant. It involves the performance of one of several types of task analysis, which is a method for describing the physical and environmental characteristics of a task along with the skills, knowledge, and capabilities required of those who perform the task. A Human Error Analysis will identify error-likely situations that can cause or lead to an accident. A Human Error Analysis can also be used to trace the cause of a given type of human error. This type of analysis can be performed in conjunction with a Human Factors Engineering Analysis, a Human Reliability Analysis, or any of several types of system analysis.

Attributes

Purpose: Identify potential human errors and their effects or identify the cause of observed human errors.

When to Use:

a. Design: Human Error Analysis can be used to identify hardware features and features of job design that are likely to produce a high rate of human error.

b. Construction: Human Error Analysis can be used to evaluate the effect of design modifications on operator performance.

c. Operation: Human Error Analysis can be used to identify the source of observed human error and to identify human errors that could result in accident event sequences.

<u>Type of Results</u>: Systematic listing of the types of errors likely
to be encountered during normal or emergency operation; listing of
factors contributing to such errors; proposed system modifications
to reduce the likelihood of such errors. Easily updated for design
changes or system/plant/training modifications.

<u>Nature of Results</u>: Qualitative. Includes identification of system
interface affected by particular errors and relative ranking of
errors based on probability of occurrence or severity of
consequences.

<u>Data Requirements</u>:

a.   Plant procedures
b.   Information from interviews of plant personnel
c.   Knowledge of plant layout/function/task allocation
d.   Control panel layout and alarm system layout.

<u>Staffing Requirements</u>: Staffing requirements may vary based on the
scope of the analysis. Generally, one analyst should be able to
perform a Human Error Analysis for a facility. The analyst should
be familiar with interviewing techniques and should have access to
the plant and to pertinent information such as procedures and
schematic drawings. The analyst should be familiar with or have
access to someone who is familiar with the consequences of various
types of human errors.

<u>Time and Cost Requirements</u>: The time and cost are proportional to
the size and number of tasks/systems/errors being analyzed. An hour
should be sufficient to conduct a rough Human Error Analysis of the
tasks associated with any given plant procedure. The time required
to identify the source of a given type of error will vary with the
complexity of the tasks involved, but it could be completed in an

hour. If the results of a single task analysis were used to identify several potential human errors, the time requirement per error would be significantly decreased. For example, performing a Human Error Evaluation and documenting a unit process made up of many complex, related tasks could be performed in a week or less. Identifying potential modifications to reduce the incidence of human errors would not add materially to the time required for a Human Error Analysis.

3. GUIDELINES FOR SELECTING HAZARD EVALUATION PROCEDURES

Selecting a hazard evaluation for a particular purpose can be diffi-cult. The different hazard evaluation procedures are different from each other in some ways and alike in others, as can be seen by comparing the Attributes for different procedures in Chapter 2. Moreover, there are many Factors that characterize the need for the hazard evaluation that influence the selection of the procedure. This section addresses those Factors and suggests an approach to the selection decision.

3.1 Factors Affecting Which Procedure is Selected

Phase of Process/Plant Development

As mentioned in a previous section, hazard evaluation should be a continuing process from process conception to plant shutdown and decommis-sioning. Each stage of process development has its own priorities for hazard evaluation, dependent mostly on achieving the best balance among:

1. Early identification to avoid costly redesign or construction modifications
2. Postponement of evaluation to await more detail
3. Avoidance of costly duplication of effort.

The best balance is usually achieved by using coarse screening evaluation procedures to identify major problems as early as possible and using more detailed and more costly procedures for more complete evaluations when the details on the final design and procedures are available. The complete and detailed evaluations made prior to startup can provide a useful baseline for evaluating the impact of any process/plant modifications that may be suggested during the operations phase.

Purpose of Hazard Evaluation

An earlier section described the Hazard Evaluation Process as a number of steps, each of which had its own purpose. The descriptions of the Attributes of the different procedures indicated which steps or purposes of the hazard evaluation process were covered by each procedure. In most cases each procedure will provide information for more than one step in the hazard evaluation process. Also, in many cases, the hazard evaluation step can be covered, to a greater or lesser extent, by more than one procedure.

Potential Consequence Levels

"Worst case" conservative estimates of consequence levels can influence the choice of hazard evaluation procedure. A potential large release of toxic materials can justify a more complete and detailed search for events and combinations of events that could cause such a release. Conversely, if there is high confidence of a low hazard level, a less exhaustive search for causes may be in order.

Complexity of Process/Plant

The degree of complexity can influence the choice of hazard evaluation procedure. A plant that incorporates several levels of protection through redundant controls, safety systems, mitigation systems, etc., needs an evaluation procedure that can identify, evaluate, and present the variety of accident event sequences that are possible. This is sometimes but not always a function of size. Simpler and smaller systems can be evaluated with simpler hazard evaluation procedures.

Familiarity With Procedures

A very well done, simple procedure will provide better results for decision making than a poorly done, more sophisticated procedure. Familiarity of staff with certain procedures is an argument for using them, provided that the procedures' limitations are completely understood.

Information and Data Requirements

Some of the procedures described in this document require more input data and information than others. If this information is not available, the results will not justify the use of those procedures. This is not as much of a problem when the procedures are used to provide qualitative results as when quantitative results are required.

Time and Cost Requirements

Time for analysis and cost of the evaluations should not be an absolute factor in the choice of hazard evaluation procedures. However, it is a factor which should be compared to the cost of risk reduction opportunities which might obviate or reduce the cost of the analysis. Also, there may be other choices, such as not modifying a plant because of the cost of evaluating the modifications, or not continuing to operate a marginal plant.

Opportunities for Risk Reduction

The costs of further analysis should always be compared to the cost of risk reduction methods that would lower the costs of analysis. Ordinarily the cost of analysis by experienced analysts will be small compared to most risk reduction approaches, but the company with little experience in hazard evaluation may, in some circumstances, find the reverse to be true.

3.2 Selection Process

In principle, one could select the most appropriate hazard evaluation procedure for a particular need by reviewing the Factors which characterize the need and the Attributes which characterize the hazard evaluation procedures and select the best match, giving due consideration to the relative priorities of the Factors and the Attributes that the situation demands. In practice, the Factors are difficult to quantify, as are the differences in the Attributes from one procedure to another; so a simpler approach is suggested.

The first step, after knowing the stage of process development, is to determine what steps of the hazard evaluation process (Figure 1-1) need to be accomplished. Knowing the purpose of the evaluation, one can then refer to Figure 3-1, a matrix relating hazard evaluation procedures to hazard evaluation process steps, to find candidate Procedures. The selection from among these candidates should then be made on the basis of their Attributes and an understanding of the Factors that are important to the choice. In general, the Factors are listed in order of priority, recognizing that particular situations can alter some of those priorities.

As an example of procedure selection, consider a process unit that is to use as input a material that is a toxic gas at standard conditions. The hazard is an unplanned release of the toxic gas to the surrounding area. An estimate of the "worst case" consequence is needed, and the Dow and Mond Indices, "What If" method, or Failure Modes, Effects and Criticality Analysis (FMECA) are the candidate methods indicated in Figure 3-1. Also, a Hazards and Operability Study (HazOp) would provide the context for an estimate. If any one of these had been used to identify the hazard it could also at least provide the context for the "worst case" estimate; i.e., the inventory(ies) available for release, the locations of the release points, and the causes of the release. The choice among these procedures should be made on the basis of the Attributes of the Procedures and the Factors characterizing the application as described above. The "worst case" consequence is thus estimated from the inventory and release location and assumed release rate. A somewhat more quantified value can be obtained from the Dow and Mond Hazard Indices if the extra effort is warranted at this point.

In this example, the estimate of "worst case" consequences is too large and the likelihood of such accident(s) is now of interest. To estimate the likelihood, the event sequence of the accident(s) leading to the "worst case" consequence should be defined. The first step is to identify the

Hazard Evaluation Procedures

Steps in Hazard Evaluation Process	Process/System Checklists	Safety Review	Relative Ranking Dow & Mond	Preliminary Hazard Analysis	"What If" Method	Hazard and Operability Study	Failure Modes Effects and Criticality Analysis	Fault Tree Analysis	Event Tree Analysis	Cause Consequence Analysis	Human Error Analysis
Identify Deviations From Good Practice	Primary Purpose	Primary Purpose	Primary Purpose								
Identify Hazards	Primary Purpose*	Primary Purpose*	Primary Purpose*	Primary Purpose	Primary Purpose	Primary Purpose	Primary Purpose	Provides Context Only			
Estimate "Worst Case" Consequences			Primary Purpose	Primary Purpose	Primary Purpose	Provides Context Only	Primary Purpose				
Identify Opportunities to Reduce Consequences			Primary Purpose	Secondary Purpose		Provides Context Only	Provides Context Only				
Identify Accident Initiating Events					Primary Purpose	Primary Purpose	Primary Purpose	Primary Purpose			Primary Purpose
Estimate Probabilities of Initiating Events						Provides Context Only	Provides Context Only	Primary Purpose		Primary Purpose	Primary Purpose
Identify Opportunities to Reduce Probabilities of Initiating Events								Primary Purpose		Primary Purpose	Primary Purpose
Identify Accident Event Sequences and Consequences					Primary Purpose			Primary Purpose	Primary Purpose	Primary Purpose	
Estimate Probabilities of Event Sequences								Primary Purpose	Primary Purpose	Primary Purpose	
Estimate Magnitude of Consequences of Event Sequences								Provides Context Only	Provides Context Only	Provides Context Only	
Identify Opportunities to Reduce Probabilities and/or Consequences of Event Sequences									Primary Purpose	Primary Purpose	Primary Purpose
Quantitative Hazard Evaluation								Primary Purpose	Primary Purpose	Primary Purpose	Primary Purpose

* Previously Recognized Hazards Only.

FIGURE 3-1. MATRIX RELATING HAZARD EVALUATION PROCEDURES TO HAZARD EVALUATION PROCESS STEPS

initiating event. The "What If" method, HazOp Study, FMECA, or Fault Tree Analysis can be used. The probability of the initiating event may then be obtained from failure rate data or human error probabilities, if available, or by using Fault Tree Analysis to relate the initiating event to its causes for which there may be failure rate or human error data or estimates.

For this example let us assume that the initiating event, failure of a coupling, dominates the accident likelihood. The failure probability of the coupling is found to be sufficiently low that the likelihood of the accident is considered to be remote. If that were not the case, alternative coupling methods would be sought.

If the system were more complex, it might be necessary to identify the entire event sequence and estimate and combine the probabilities of several of the events in order to estimate the likelihood of the accident. The "What If" method, Fault Tree Analysis, Event Tree Analysis, and Cause-Consequence Analysis are candidate procedures for identifying the event sequences, with the latter three being more structured. Again, the choice is made on the basis of the Attributes and Factors as described. The latter three procedures also provide the structure for combining event probabilities and, if data are available, a quantitative result may be obtained.

4. <u>GUIDELINES FOR USING HAZARD EVALUATION PROCEDURES</u>

This chapter provides guidelines for performing hazard evaluations using each of the following procedures:

<u>Procedure</u>	<u>Page</u>
Process/System Checklists............................	4-2
Safety Review.......................................	4-9
Relative Ranking--Dow and Mond Hazard Indices........	4-15
Preliminary Hazard Analysis.........................	4-20
"What If" Analysis..................................	4-25
Hazard and Operability (HazOp) Studies..............	4-33
Failure Modes, Effects, and Criticality Analysis.....	4-53
Fault Tree Analysis.................................	4-61
Event Tree Analysis.................................	4-82
Cause-Consequence Analysis..........................	4-93
Human Error Analysis................................	4-97

Each guideline differs somewhat in format because of the differing focus of each hazard evaluation procedure.

4.1 Process/System Checklists

General Description

A process or system checklist can be applied to evaluating equipment, materials, or procedures. Also, the checklist can be used during any stage of a project to guide the user through common hazards by using standard procedures.

The checklist should be prepared by an experienced engineer who is familiar with the general plant operation and the company's standard procedures. Once the checklist has been prepared, it can then be applied by less experienced engineers and reviewed by a manager or staff engineer who has approval authority for the appropriate subsequent action.

A checklist will generally produce compliance with the minimal standards and identify areas that require further evaluation.

Guidelines for Using Procedure

Examples of checklists for each of five phases of a project's life (design, construction, start-up, operation, and shutdown) are provided below. These checklists are necessarily general, with a discussion of many of the elements to be checked. To be most useful, the checklists should be specifically tailored for an individual company, plant, or product.

These brief checklists are for illustrative purposes only. Examples of more complete checklists are shown in Appendix A.

1. Design

During the design effort, the extent of hazard evaluation should be consistent with the preliminary or final design. The following discussion is appropriate for a final design and is more extensive than what would normally be needed for a preliminary design effort.

<u>Materials</u>. Review the characteristics of all process materials--
raw materials, catalysts, intermediate products, and final products. Obtain
detailed data on these materials such as:

- Flammability
  - What is the autoignition temperature?
  - What is the flash point?
  - How can a fire be extinguished?
- Explosivity
  - What are the upper and lower explosive limits?
  - Does the material decompose explosively?
- Toxicity
  - What are the breathing exposure limits (e.g., Threshold Limit
    Values, Immediately Dangerous to Life and Health, etc.)?
  - What personal protective equipment is needed?
- Corrosivity and Compatibility
  - Is the material strongly acidic or basic?
  - Are special materials required to contain it?
  - What personal protective equipment is needed?
- Waste Disposal
  - Can gases be released directly to the atmosphere?
  - Can liquids be released directly to a sewer?
  - Is a supply of inert gas available for purging equipment?
  - How would a leak be detected?
- Storage
  - Will any spill be contained?
  - Is this material stable in storage?
- Static Electricity
  - Is bonding or grounding of equipment needed?
  - What is the conductivity of the materials, and how likely are
    they to accumulate static?

- Reactivity
  - Critical temperature for auto reaction?
  - Reactivity with other components including intermediates?
  - Effect of impurities?

Equipment. Review the process flow sheet or equipment list to identify the hazards associated with each piece of equipment.

- Design Specifications
  - Are design margins (or safety factors) clearly stated for temperature, pressure, flow, level, or other process variable?
  - Are the materials of construction compatible with the process stream?
  - Is the equipment subject to stress corrosion cracking (e.g., stainless steel)?
- Pressure Relief
  - Have safety valves been sized for all conditions (i.e., fire, loss of cooling, closed downstream valve)?
  - Do pipelines need pressure relief for thermal expansion?
  - Is there risk of the vents plugging?
- Plant Arrangement
  - Has adequate spacing been provided between pieces of equipment?
  - Are flame arrestors needed for equipment vents?
  - Do any pieces of equipment need remotely operated valves?
- Electrical Equipment
  - Does all equipment comply with the electrical classification?

Procedures. During the course of the design, the proper procedures for handling startup, shutdown, or emergency situations should be addressed.

- How should the plant operators and instrumentation and control systems react to the following contingencies?
  - Fire
  - Gas release
  - Electrical power failure

- Cooling water failure
- Steam failure
- Loss of instrumentation (instrument air or electrical power)
- Loss of inert gas supply
- Do any plant interlocks have to be bypassed for plant startup or shutdown?
- Does the design consider a natural disaster such as earthquake, flood, or tornado?

2. Construction

During the construction period, several critical actions are required to avoid problems with adjacent facilities or in future operation.

Materials
- Has an authorized individual checked that the material received conforms to the design specifications?
- Have spare parts been ordered?

Equipment
- Has a hydrostatic test or other acceptance test been witnessed by an authorized individual?
- Are battery limit block valves accessible and clearly identified?

Procedures
- Are skilled craftsman being used?
- Is good housekeeping being practiced?
- Has an orderly transfer to Operations been planned?

3. Startups

Every possible effort should be made to simplify problems and actions during this hectic and critical period. This is an especially vulnerable time that requires thoughtful, deliberate action to avoid errors.

Materials

- Has there been prior planning for raw materials, utilities, and operating supplies to simplify any required action during startup?
- Have all suppliers been notified of startup plans?
- Have contingency plans been made for disposal of off-spec material?

Equipment

- Has all equipment been purged of air (if required)?
- Have all blinds been removed?
- Are all valves in their correct position?
- Have instrument/interlock checks been completed?
- Have all critical pieces of equipment been clearly identified?
- Has the final inspection/acceptance of all equipment been completed?
- Does instrumentation fail safe?

Procedures

- Have startup and operating procedures been prepared in advance?
- Is operator training complete?
- Has a walk-through of the startup been completed?
- Have the startup plan and schedule been clearly communicated?
- Are emergency procedures complete?

4. Operation

Once a plant has been operating for a long time, there is a tendency to become complacent about hazards. Plant operations staff must remain diligent to identify and minimize hazards.

Materials

- Do all raw materials continue to conform to specifications?
- Is each receipt of material checked for identification?
- What routine tests are needed to support plant operations?

- Does the operating staff have access to Material Safety Data Sheets?
- Is fire fighting and safety equipment properly located and maintained?

Equipment

- Has all equipment been inspected as scheduled?
- Have pressure relief valves been inspected as scheduled?
- Have safety systems and interlocks been tested at appropriate times?
- Are the proper maintenance materials (spare parts) available?

Procedures

- Are the Operating Procedures current?
- Are the operators following the Operating Procedures?
- Are new operating staff trained properly?
- Are experienced operating staff kept up to date on all Operating Procedures?
- How are communications handled at shift change?
- Is the Operating Logbook being used properly?
- Is housekeeping acceptable?
- Are electrical tag-out procedures being used?
- Are work permits being used?
- Is gas testing being performed diligently?

5. Shutdown

This phase of a project or plant is usually neglected because people tend to look forward to the next project. Several hazards can be overlooked at this point unless diligence is maintained.

Materials

- Has the inventory of all chemicals been removed?
- Has all equipment been purged or flushed with inert material?

Equipment

- Is the equipment, including piping, gas-free?
- Have entry barriers been located at necessary points?

Procedures

- Has a shutdown plan or schedule been communicated to the appropriate staff?

4.2 Safety Review

General Description

An intensive plant inspection, often called a "Safety Review", is intended to identify plant conditions or operating procedures that could lead to accidents and significant losses in life or property. Such an inspection may also be called a process safety review, a loss prevention review, etc. It is often done a year after plant startup and repeated at several year intervals. This comprehensive inspection, usually performed by a team with varied backgrounds and responsibilities, is intended to complement other plant efforts such as routine visual inspections. A team of 2-5 people may participate for a week or more in conducting a complete Safety Review. There are many variations of Safety Reviews following the general ideas described below.

Interviews with plant staff (operators, maintenance staff, engineers, management, and safety staff) are needed so that the facility can be examined thoroughly from many perspectives. The inspection should seek to identify major risk situations. General housekeeping and personnel attitude are not the primary objectives but may be significant indicators of where to look for real problems or needed improvements. Various hazard evaluation techniques, such as checklists, "what-if" questions, and raw material evaluations, can be used during the inspection.

In preparation for a Safety Review, the following should be done:

- Assemble a detailed description of the plant (e.g., plot plans, P&ID's) and procedures (e.g., operating, maintenance, emergency)
- Assemble all applicable codes and standards
- Schedule interviews with specific individuals
- Request that records be available (e.g., personnel injury, hazardous incident reports, pressure vessel inspection, pressure relief valve testing)
- Plan a wrap-up visit with plant manager or appropriate alternate.

At the end of the inspection, specific actions are recommended, with designated responsibilities and completion dates. A follow-up evaluation, or re-inspection, should be planned to verify the acceptability of the corrective action.

Guidelines for Using Procedure

Concept. Periodic inspections of an operating plant help to ensure that implemented safety/risk management programs meet original expectations and standards. Inspections keep operating personnel alert to the process hazards as they respond to questions from a knowledgeable inspection team. The review seeks to identify operating procedures that need to be revised and new equipment or process changes that may have introduced new hazards, and to evaluate the adequacy of equipment maintenance or replacement. A Safety Review often identifies opportunities for new technology to be applied to an existing hazard to reduce the risk if necessary.

The complete inspection is directed toward every aspect of the operating facility. It can also be used for pilot plants, laboratories, storage facilities, or supporting services. The inspection should include a process review and should address all plant equipment, instrumentation, associated utilities, environmental protection facilities, maintenance areas, and fire, safety, training, and security services. An evaluation of the know-ledge and training of the staff of each group is included. The cooperation of each individual and group is necessary to improve the overall safety and performance of the plant. Therefore, people need to understand the benefit of conducting Safety Reviews so that they can participate in and endorse the recommendations.

Procedure. Interviews with specific individuals should be scheduled, including an exit debriefing with the plant manager. Records should be requested for such things as personnel injuries, hazardous incidents, pressure vessel inspections, pressure relief valve testing, critical instrument checks, and chemical reactivity data.

The inspection schedule should start with a general orientation tour of the plant and then move to specific interviews. Any part of the inspection that involves outdoor work should be scheduled early in the survey to allow rescheduling for inclement weather.

The inspection team obtains and reviews current copies of plant drawings (P&ID, flowcharts) and operating, maintenance, and emergency procedures. These serve as the basis for review of the process hazards and for discussions with operating personnel. Because most injuries are caused by failure to follow established procedures, an effort should be made to determine whether the operating staff follow their procedures. This evaluation can extend to the control over maintenance activities such as routine equipment repair, welding, vessel entry, electrical lock-out, or equipment testing. The observation of people as they perform their daily tasks provides evidence of knowledge of procedures and compliance. Questions can further clarify why and how some action is being handled.

Useful information can be gained by conducting one or more mock emergencies in which all staff are asked to respond as they would in a real emergency. The participating staff walk through the exercise, explaining to the inspection team what they would do in sequence and why and how they expect the system to react to their action. The inspection team observes the exercise and, after the exercise, provides a critique to the participants. The section of the plant's Operating Procedures covering emergencies is used to identify the expected performance.

Equipment inspection requires visual or diagnostic evaluation plus a review of plant records. This effort is generally reserved for the most critical equipment. Some questions that might be addressed are: Is the equipment in good general condition? Are the pressure relief or other safety devices installed properly, well maintained, and properly identified? Do plant records show the history of testing of the equipment or the safety device(s)? Has a pressure vessel been hydrotested periodically? For equipment that handles corrosive or erosive materials, have metal-wall-thickness measurements been taken at frequent intervals? Does the plant have trained inspectors whose recommendations for repair or replacement are accepted by management?

Critical instruments and safety interlock systems deserve special attention during a Safety Review. Automatic controls or emergency shutdown systems need periodic testing to verify that the controls work as intended. If possible, it is prudent to schedule a planned plant shutdown (outage) so that the safety systems are activated and tested under close scrutiny. If this is not possible, then one must depend upon functional checks. Bypassing safety controls is a serious matter that should be checked and evaluated.

An inspection of fire and safety equipment is appropriate to ensure that it is well maintained and that the staff are properly trained to use it. Simple questions such as "Can the fire equipment reach the top level of a structure?" should be checked. If there are sprinkler, dry chemical, or foam fire suppression systems, are they checked on a periodic basis? Emergency response plans should be reviewed for completeness and tested if possible.

When the inspection is completed, a report is prepared with specific recommended actions. Justification for the recommendations is provided in the report, and both the favorable and unfavorable impressions that have been formed are summarized. Findings and recommendations are reviewed with appropriate plant management. Any follow-up action should be planned at this time.

Results. A report of findings and recommendations is the final product of the inspection team. Implementation of corrective actions should remain the responsibility of plant management. Any follow-up effort will depend upon the situation, but certainly some confirmation is needed that the recommendations have been heeded.

The major objective of the inspection is to make the plant safer, both by maintaining the staff's awareness of hazards and by correcting situations (equipment or procedures) that needed corrective action.

From the perspective of corporate management, such inspections indicate the safety and risk management performance of various operating plants. This indication is necessary to judge the need for management attention and corporate resources.

Example. In a major petrochemical complex, one of the operating units was approximately 30 years old. A business evaluation indicated that this unit should continue to operate for another 15-20 years if the equipment was serviceable. The plant had been well maintained, and a thorough plant inspection indicated that the equipment was serviceable for an additional 15 years. However, the comprehensive equipment inspection identified several problems that needed additional evaluation:

- The unit capacity had been increased during its life through numerous improvements and de-bottlenecking projects, yet the unit's safety valves and flare headers had never been resized.
- The unit instrumentation was the original pneumatic controls, dating back to 1940, without any safety interlocks or shutdowns.
- The corporate standards for equipment spacing had changed since the unit was originally built, and by the new standard, there were numerous violations of the recommended safe equipment spacing.

Each of these problems was addressed by the company:

- A thorough design review was conducted of the pressure relief valves and flare headers. As a result of this study, additional safety valves and a new flare header were installed.
- The design review expanded to consider the instrumentation and equipment spacing. A modern control system was recommended and installed. Not only was this system more reliable, but it also provided some safety interlocks that did not exist originally.
- The equipment spacing problem was considerably more difficult. Certain spacing requirements could not be met with the existing unit and the adjacent facilities (flare, furnaces, and control room), so a heat-activated sprinkler system was installed to address the possibility of a fire on this unit. Spacing was also a consideration in a subsequent decision to build a new consolidated control room farther away from this unit.

The company accepted the considerable cost of these efforts because the improved plant could clearly be expected to operate safely for another 15 years.

4.3 Relative Ranking Techniques--Dow and Mond
Hazard Indices

General Description

The Dow and Mond Indices provide a direct and easy method for quickly estimating the risks in a process plant. The methods assign penalties and credits based on plant features. Penalties are assigned to process materials and conditions that can contribute to an accident. Credits are assigned to plant safety features that can mitigate the effects of an accident. These penalties and credits are combined to derive an index that is a relative ranking of the plant risk.

Both the Dow and Mond ranking methods address risk in individual process units. The two methods are closely related; the Mond Index was developed as an extension of the Dow Index. The primary difference in the two methods is that the Mond Index specifically addresses material toxicity in addition to flammability and reactivity in assigning material factors to the process units. The Dow Index may actually be easier to use because of the extensive use of tables and graphs in place of traditional equations, but both methods use the same basic calculational method. The fifth edition of the Dow guide includes methods for estimating outage time and business interruption costs; these methods can be applied with either index.

This discussion does not attempt to reproduce the Technical Guides for these methods, and the reader is referred to these Guides for the technical details that are necessary to apply the techniques.

Guidelines for Using Procedure

The relative ranking methods consist of seven general steps:
1. Identify on the plot plan those "process units" that would have the greatest effect or contribute the most to a fire, explosion, or release of toxic material.
2. Determine the Material Factor (MF) for each unit.

3. Evaluate the appropriate Contributing Hazard Factors, considering fire, explosion, and toxicity.

4. Calculate the Unit Hazard Factor and Damage Factor for each unit.

5. Determine the Fire and Explosion Index and Area of Exposure for each unit.

6. Calculate Maximum Probable Property Damage (MPPD), Base and Actual.

7. Evaluate Maximum Probable Days Outage (MPDO) and Business Interruption (BI) costs.

A discussion of each of these steps follows.

1. <u>Identify on the plot plan those "process units" that would have the greatest effect or contribute the most to a fire, explosion, or release of toxic material</u>. Selecting the process units provides the focus of the analysis. A process unit is defined as any primary item of process equipment, such as a pump, reactor, compressor, or storage tank. Some analyses might need to include all units within a process area, but normally certain units are prime candidates for the analysis because of inventory, reaction type, or process conditions. The objective is to select the units that would be most likely to be involved in a fire, explosion, or toxic material release.

2. <u>Determine the Material Factor (MF) for Each Unit</u>. The Material Factor measures the intensity of the potential energy release in the process unit. The MF considers the flammability, reactivity, and toxicity of the material in the unit. MFs in the Dow Guide are denoted by a number from 1 to 40. The Guide lists MF values for 300 materials, plus instructions for determining the MF of a material not included in the list. The Mond Guide also provides the procedures needed to include material toxicity in the evaluation.

3. <u>Evaluate the Appropriate Contributing Hazard Factors</u>. The contributing hazards are of two types: general process hazards and special process hazards. Each of these could be a contributing factor to an incident that could result in a fire or explosion. General process hazards increase

the magnitude of the incident, and special process hazards increase the probability of an incident.

General process hazards (designated "F_1") include exothermic reactions; endothermic reactions; material handling and transfer; enclosed process units; inadequate access; and drainage.

Special process hazards (designated "F_2") include process temperature; low pressure (sub-atmospheric); operation in or near flammable range; dust explosion; relief pressure; low temperature; quantity of flammable material; corrosion and erosion; leakage in joints and packing; use of fired heaters; hot oil heat exchange system; and rotating equipment, such as pumps and compressors.

The general and special process hazards are the sum of all the individual penalties for each category plus a base factor of 1.0.

4.   Calculate the Unit Hazard Factor and Damage Factor for Each Unit.   The Unit Hazard Factor (F_3) is simply the product of the general and special process hazards (F_1 and F_2) identified in the previous step. The Unit Hazard Factor is combined with the Material Factor (from Step 2) to determine the Damage Factor. The Dow Guide provides a graph that allows easy determination of the Damage Factor from its inputs. The Damage Factor is a measure of the probable relative damage exposure.

5.   Determine the Fire and Explosion Index and Area of Exposure for Each Unit. The Fire and Explosion (F&E) Index measures the probable damage that may result in the process plant. The F&E Index is calculated as the product of the Unit Hazard Factor and the Material Factor. Since a fire and/or explosion event includes blast effects and fire exposure from the original release, as well as possible propagation of the fire, the F&E Index is related to an Area of Exposure. The Area of Exposure is defined as a circular area around the process unit. The Dow Guide provides a graph that allows easy determination of the radius of this circle from the F&E Index. The Dow F&E Index can also be used as a general rule for specifying the degree of hazard at the unit. The guideline for this rule is:

Dow F&E INDEX RANGE	DEGREE OF HAZARD
1 - 60	Light
61 - 96	Moderate
97 - 127	Intermediate
128 - 158	Heavy
159 - Up	Severe

6. <u>Calculate Maximum Probable Property Damage (MPPD), Base and Actual</u>. The Base MPPD is calculated from the replacement value of the equipment within the Area of Exposure (Step 5). The calculation is based on the equipment original cost, an allowance for items not subject to loss (site preparation, design engineering, etc.), and an escalation factor if needed. For calculating the Base MPPD, the Area of Exposure can be modified if blast- or fire-resistant walls provide protection to equipment. Product Inventory should also be included in the Base MPPD.

Actual MPPD is derived from the Base MPPD by applying Loss Control Credit Factors. The Loss Control Credit Factors account for the mitigating effects of basic safety design features of the plant. The procedure defines three categories of loss control features.

A. Process Control, such as emergency power, explosion control, emergency shutdown, computer control, inert gas, operation instructions, and reactive chemical review.

B. Material Isolation, such as remote control valves, dump/blowdown, drainage, and interlock.

C. Fire Protection, such as leak detection, structural steel, buried tanks, water supply, sprinkler systems, water curtains, foam, hand extinguishers or monitor guns, and cable protection.

The credit factor for each category is the product of the credits for each item within the category, and the total credit is obtained from a graph using the product of the three category credits. Actual MPPD is calculated as the product of the total credit and the Base MPPD.

7. <u>Evaluate Maximum Probable Days Outage (MPDO) and Business Interruption (BI) Costs</u>. The consequence of an incident can be estimated by the MPDO and BI costs. This calculation considers:

A. Cost of repairing or replacing damaged equipment (property damage)

B. Loss of production, stated as the Value of Product Manufactured (VPM).

The Dow Guide provides a graph for calculating the MPDO using the Actual MPPD (Step 6). This calculation is based on experience data from 137 incidents and includes a 70-percent-probability range for the MPDO. The BI costs can then be calculated from the MPDO and the VPM.

<u>References</u>

1. Dow Chemical Company, Fire and Explosion Index, Hazard Classification Guide, 5th Edition, 1981.

2. Lewis, D. J., "The MOND Fire and Explosion Index Applied to Plant Layout and Spacing", 13th Loss Prevention Symposium, 1979.

4.4 Preliminary Hazard Analysis

General Description

The Preliminary Hazard Analysis (PHA) is based on the technique as defined and used by the military in their system safety programs (Department of Defense, 1984). Many chemical companies have a similar procedure in place, perhaps under a different name. This analysis has been found to be very cost effective in the development phase of all hazardous military systems, including process plants. It is also possible to use such a procedure to prioritize other more detailed hazard identification methods for use at other times during a plant's lifetime.

A PHA is designed to be used during the concept, early development, or siting phase of a process plant to determine the hazards that exist. A PHA does not preclude the need for further hazard assessment; instead, it is a precursor to subsequent hazard analyses. The principal advantages of the PHA are: (1) early identification and awareness of potential hazards by the design team; and (2) identification and/or development of guidelines and criteria for the process development team to follow. Thus, as the project develops, the principal hazards can be eliminated, minimized, or controlled almost from the start.

The PHA is accomplished by listing the hazards associated with the elements of the system as defined in the concept or early design stage. Plant elements which may be definable at this stage include:

- Raw materials, intermediate and final products and their reactivity
- Plant equipment
- Interface between components
- Operating environment
- Operations (test, maintenance, emergency procedures, etc.)
- Facility
- Safety equipment.

As each hazard is identified, the potential causes, effects, and severity of accidents and the possible corrective and/or preventive measures are also listed. Completeness is achieved by utilizing past experience from as many different sources as possible. These sources include hazard studies of similar facilities, operating experience from similar facilities, and hazard listings, such as those included in Appendix A of this document.

Guidelines for Using Procedure

The Preliminary Hazard Analysis consists of the following basic steps:

1. Gather needed information
2. Carry out Preliminary Hazard Analysis
3. Record results.

1. Gather Needed Information

The PHA requires gathering, first of all, information available on the subject plant (or system) and then relevant information from past experience with any similar plant or even from a plant which has a different process but uses similar equipment and materials.

Because the PHA is specifically intended for early hazard identification, information on the plant may be sparse. At the point in the design development where a PHA is effective, the minimal information available will include the process concept. Thus, the basic chemicals and reactions involved should be known, as well as the major types of equipment, especially long-lead-time or special equipment items (e.g., vessels, heat exchangers, and facility construction). In addition to the components of the plant, the operational goals of the plant and basic performance requirements are useful in defining the context for hazards and the environment in which the plant will operate.

It is very useful to determine what past experience is available with the chemicals and/or the process concept involved. Any problems known through past experience can help in the PHA for the subject plant.

2. Carry Out Preliminary Hazard Analysis

The process of accomplishing the PHA is to identify the hazards, potential initiating events, and other events which could result in an undesired consequence. The analyst(s) should also identify design criteria or alternatives that could eliminate or reduce those hazards judged to lead to an excessive risk level. Obviously, some experience is required in making such judgments. In conducting the PHA the following items should be considered:

a. Hazardous plant equipment and materials (e.g., fuels, highly reactive chemicals, toxic substances, high pressure systems, and other energy storage systems)

b. Safety-related interfaces between plant equipment items and materials (e.g., material interactions, fire/explosion initiation and propagation, and control/shutdown systems)

c. Environmental factors that may influence the plant equipment and materials (e.g., earthquake, vibration, extreme temperatures, electrostatic discharge, and humidity)

d. Operating, testing, maintenance and emergency procedures (e.g., human error importance, operator functions to be accomplished, equipment layout/accessability, and personnel accident protection)

e. Facilities support (e.g., storage, testing equipment, training, and utilities)

f. Safety-related equipment (e.g., mitigating systems, redundancy, fire suppression, and personal protection equipment).

As an example, consider a concept which will use liquid H_2S as feed stock. The PHA analyst has only the information that this product will be used in the process--no other design details. The analyst recognizes that H_2S is toxic and identifies its release as a hazard. The analyst considers causes for such a release; for example:

- The pressurized storage cylinder leaks or ruptures
- The process does not use up all of the H_2S
- The H_2S process supply lines leak/rupture
- A leak occurs during receipt of H_2S at the plant.

The analyst then determines the effect of these causes; in this case, fatalities could result from larger releases. The next task is to provide guidance and criteria for designers to use in designing the plant, recognizing each of the significant, potential release mechanisms. For example, for the first item, storage cylinder leak, the analyst might lead the designer to:

a. Consider a process which stores alternative, less toxic materials that can generate H_2S as needed

b. Provide a plant warning system

c. Minimize on-site storage of H_2S, without excessive delivery/handling (e.g., storage of a two-weeks' to one-month's production requirement)

d. Develop a procedure for storage cylinder inspection with a human factors engineer

e. Consider a cylinder enclosure with a deluge system triggered by leak detector

f. Locate the storage cylinder for easy access for delivery, but away from other plant traffic

g. Suggest development of a training program for all employees on H_2S effects and emergency procedures, to be presented before startup to all employees and subsequently to all new employees; consider a similar program for plant neighbors.

3.   Record Results

        The PHA results are conveniently recorded on a form which displays
the hazards identified, the cause, the potential consequence, and any
identifiable corrective or preventive measure. Figure 4-1 displays an
example. The result is normally provided to the project manager and the plant
design staff.

Hazard	Cause(s)	Major Effects	Corrective/Preventive Measures
Toxic release	1) H_2S storage cylinder leak	Potential for fatalities from large release	(a) Provide warning system (b) Minimize on-site storage (c) Develop procedure for cylinder inspection with human factors engineer
	2) H_2S not used up in process	(as above)	(a) Design excess H_2S collection system and burn off excess (b) Control system design to detect excess H_2S and shut down process (c) Develop procedures to include availability of excess-burnoff system prior to plant startup

FIGURE 4-1.   SAMPLE PRELIMINARY HAZARD ANALYSIS WORKSHEET

References

1.   Hammer, Willie, Handbook of System and Product Safety, Prentice Hall, Inc.,
     1972.

2.   Department of Defense, Military Standard System Safety Program Requirements
     MIL-STD-882B, Washington, D.C.

4.5 "What If" Analysis

Description

The "What If" analysis has had limited reported usage, and information on it in the open literature is sparse. However it is frequently referred to in the industry, may be the method most in use, and enjoys a good reputation among those skilled in its use.

The concept of a "What If" analysis is to conduct a thorough and systematic examination of a process or operation by asking questions that begin with "What If...". This examination can include buildings, power systems, raw materials, products, storage, material handling, in-plant environment, operating procedures, work practices, management practices, plant security, etc. This is a powerful procedure if the staff is experienced; otherwise the results are likely to be incomplete.

The formulation of the exact questions is left up to those conducting the examination, unless questions have been formulated by the plant in a previous review. The questioning usually starts at the input to the process and follows the flow of the process. Alternatively, the questioning can center on a particular consequence category, e.g., personnel safety or public safety. Usually a small group of two or three experts conducts the examination and reports the findings. The findings are usually accident event sequences that result from the "What If" questions. The questions essentially suggest an initiating event, and perhaps a failure from which an undesirable event sequence could occur. For example, a question might be:

"What if the raw material is the wrong concentration?"

The team would then attempt to determine how the process would respond; for example:

"If the concentration of acid were doubled, the reaction could not be controlled and a rapid exotherm would result."

The team might then recommend, for example, installing an emergency shutdown system or taking special precautions when loading the raw material.

The questions and the answers, including hazards, consequences, and solutions of importance, are all recorded. The study is completed when the team (or teams, if several areas are being investigated) completes its recording and reports the results to the responsible individual. The recommendations are then reviewed with the appropriate level of management.

Guidelines for Using the Procedure

The "What If" analysis includes the following steps:

1. Define study boundaries
2. Gather needed information
3. Define the team
4. Conduct the reviews
5. Record results.

1. Define Study Boundaries

There are two types of boundaries in a "What If" study: the physical system boundaries and the consequence category being investigated.

These two boundaries are closely related; that is, the consequences to be considered (public safety, plant damage, etc.) will define what portions of the plant to consider. Thus, it is first necessary to define the consequence categories to be considered. In general, these can be divided into the broad categories of public risk, worker risk, and economic risk. These can be further divided into specific, more detailed areas, and even types of accident event sequences. For example, economic risk could be defined, for a specific plant, to be "fires leading to loss of production for more than 3 days". Considerable care should be exercised in becoming this specific; although specific problem definitions aid the team(s) in conducting the "What If" procedure, they also restrict the team to these areas, possibly causing other important accident event sequences to be overlooked.

After the consequence categories are defined for the study, the physical boundaries of the study can be defined. The purpose of defining these boundaries is to keep the team focused on portions of the plant in which the consequence of concern could occur. Care must be taken in defining these boundaries because very often there are interactions between parts of a plant, some of which may not be hazardous in themselves (relative to the consequence) but which may cause some other portion of the plant to perform abnormally and hence result in an accident event sequence. The boundaries should include any interacting portion of the plant.

2. Gather Needed Information

After the study boundaries are defined, the needed information is gathered for the teams. This information can be expected to include the items shown in Table 4-1. It is important that all the information be available to the review teams to allow the examination process to continue without interruption.

If an operating plant is being reviewed, it is possible that the team(s) may want to question the plant operating, maintenance, or other staff. Thus, before the review begins, arrangements should be made with the plant. In addition, if the team is off-site, they may want to visit the plant to get a better idea of how the plant is "put together" and how it operates.

The last part of the information gathering is the preparation of the questions. If this review is an update of a previous review or an examination of a plant change, previous questions can be used. For a new plant or a first-time application, the questions should be developed as much as possible before the team meetings. Other types of studies described in this document (such as HazOp and Checklists) can aid in this process. It will also be necessary for the "What If" team members to add questions as they conduct the review.

TABLE 4-1. TYPICAL INFORMATION NEEDED FOR "WHAT IF" TECHNIQUE

I. Process Flow Sheets

   1. Operating Conditions

      a. Process materials used, including physical properties
      b. Process chemistry and thermodynamics

   2. Equipment Description

II. Plot Plan

III. Process and Instrumentation Drawings

   1. Controls

      a. Continuous monitoring devices
      b. Alarms and their function

   2. Instrumentation

      a. Charts
      b. Gauges
      c. Monitors

IV. Operations

   1. Responsibilities and duties of operating personnel
   2. Communications systems
   3. Procedures

      a. Preventive maintenance
      b. Hot work permits
      c. Vessel entry
      d. Lock-out/Tag out
      e. Emergency

3. Define the Team

      Teams of two or three members are defined for each identified investigation area. The team should include:

a. Expertise in the consequence category

b. Knowledge of the plant/process

c. Experience in general hazard evaluation.

This team could include a first-line supervisor of manufacturing, a first-line supervisor of operations, and staff from operations, maintenance, and engineering.

4. Conduct the Reviews

The review should begin with a basic explanation of the process, using the gathered plant information. This orientation should be given by plant staff who have overall plant and process knowledge plus expertise relevant to the team's investigation area. The presentation should also describe the plant's safety precautions, safety equipment, and health control procedures.

When the team has been given the information, they conduct the review. The team should not limit themselves to the prepared "What If" questions, but rather should use their combined skills and team interaction to make sure that the investigation is thorough. The time needed to complete a review depends on size of the plant. The team should not be rushed and should not work too many hours consecutively. Ideally, a team should meet 4 to 6 hours per day, every other day.

The team generally proceeds from the inputs to the process to the output. They address each prepared "What If" question, answer the question (or indicate a need for more information), and identify the hazard, potential consequences, and solutions. In the process, they add new "What If" questions that may become apparent during the analysis.

Consider, as a simple example, the continuous process shown in Figure 4-2. In this process, phosphoric acid and ammonia are mixed, and a nonhazardous product, diammonium phosphate (DAP), results if the reaction of ammonia is complete. If the proportion of phosphoric acid is reduced, the reaction is incomplete, and ammonia is evolved. Reduction in the quantity of ammonia available to the reactor results in a safe but undesirable product.

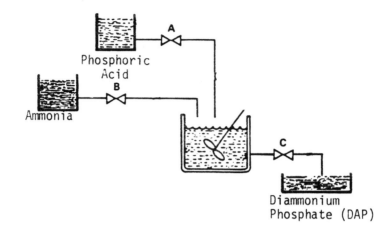

FIGURE 4-2.   CONTINUOUS PROCESS EXAMPLE FOR "WHAT IF" TECHNIQUE

TABLE 4-2.   "WHAT IF" QUESTIONS

"What If"

1. Wrong product is delivered instead of phosphoric acid
2. Phosphoric acid is wrong concentration
3. Phosphoric acid is contaminated
4. Valve A is closed or plugged
5. Too high a proportion of ammonia is supplied to reactor
6. Vessel agitation stops
7. Valve C is closed

The team is assigned to investigate "Personnel Hazards From the Reaction". Table 4-2 is a list of "What If" questions prepared for their consideration.

The team addresses the first question and would probably consider what other products could be mixed with ammonia that would produce a hazard. If one or more such products are known, then their availability at the plant is noted, along with the possibility that the vendor could deliver something marked phosphoric acid, but which is another product of that vendor. The hazards of the potential wrong product combinations are identified if they endanger the plant employees. Remedies are suggested to guard against the wrong products being used in place of phosphoric acid. The team continues through the questions in this manner until they reach the output of the process. In this case, the team noted that the process is located in an add-on block building at the plant and that extreme weather conditions may be outside of the range of the heating or air conditioning system. Thus, they add two (perhaps more) questions.

"What If" the outside temperature is -20°F?
"What If" the outside temperature is 100°F?

5. Record the Results

As with any study, reporting is the key to transforming the teams' findings into measures for hazard elimination or reduction. Figure 4-3 is an example "What If" reporting mechanism. Such a figure makes the reporting easier and more organized. These results should be reviewed with the plant manager and safety officer to make sure that the findings are transmitted to those ultimately responsible for any action.

What If	Consequence/Hazard	Recommendation
Wrong product is delivered instead of phosphoric acid	None likely	
Phosphoric acid is wrong concentration	Ammonia is not used up and is released to work area	Verify phosphoric acid concentration after filling vat prior to operation.
Phosphoric acid is contaminated	None likely	
Valve A is closed or plugged	Ammonia unreacted, released to work area	Alarm/shutoff of ammonia (valve B) on low flow from valve A into reactor.
Too high a proportion of ammonia is supplied to reactor	Excess ammonia released to work area	Alarm/shutoff of ammonia (valve B) on high flow from valve B into reactor
. . .	. . .	. . .

FIGURE 4-3. SAMPLE "WHAT IF" WORKSHEET FOR DAP PLANT

4.6 Hazard and Operability (HazOp) Studies

General Description

Background. A HazOp study identifies hazards and operability problems. The concept involves investigating how the plant might deviate from the design intent. If, in the process of identifying problems during a HazOp study, a solution becomes apparent, it is recorded as part of the HazOp result; however, care must be taken to avoid trying to find solutions which are not so apparent, because the prime objective for the HazOp is problem identification.

Although the HazOp study was developed to supplement experience-based practices when a new design or technology is involved, its use has expanded to almost all phases of a plant's life. HazOp is based on the principle that several experts with different backgrounds can interact and identify more problems when working together than when working separately and combining their results.

The "Guide-Word" HazOp is the most well known of the HazOps; however, several specializations of this basic method have been developed. These specializations will be discussed as modifications of the Guide-Word approach, but they are not to be regarded as less useful than the Guide-Word approach. Indeed, in many situations these variations may be more effective than the Guide-Word approach.

Concept. The HazOp concept is to review the plant in a series of meetings, during which a multidisciplinary team methodically "brainstorms" the plant design, following the structure provided by the guide words and the team leader's experience.

The primary advantage of this brainstorming is that it stimulates creativity and generates ideas. This creativity results from the interaction of the team and their diverse backgrounds. Consequently the process requires that all team members participate (quantity breeds quality in this case), and team members must refrain from criticizing each other to the point that members hesitate to suggest ideas.

The team focuses on specific points of the design (called "study nodes"), one at a time. At each of these study nodes, deviations in the process parameters are examined using the guide words. The guide words are used to ensure that the design is explored in every conceivable way. Thus the team must identify a fairly large number of deviations, each of which must then be considered so that their potential causes and consequences can be identified.

The best time to conduct a HazOp is when the design is fairly firm. At this point, the design is well enough defined to allow meaningful answers to the questions raised in the HazOp process. Also, at this point it is still possible to change the design without a major cost. However, HazOps can be done at any stage after the design is nearly firm. For example, many older plants are upgrading their control and instrumentation systems. There is a natural relationship between the HazOp deviation approach and the usual control system design philosophy of driving deviations to zero; thus it is very effective to examine a plant as soon as the control system redesign is firm.

The success or failure of the HazOp depends on several factors:

- The completeness and accuracy of drawings and other data used as a basis for the study
- The technical skills and insights of the team
- The ability of the team to use the approach as an aid to their imagination in visualizing deviations, causes, and consequences
- The ability of the team to concentrate on the more serious hazards which are identified.

The process is systematic and it is helpful to define the terms that are used:

a. STUDY NODES- The locations (on piping and instrumentation drawings and procedures) at which the process parameters are investigated for deviations.

b. INTENTION- The intention defines how the plant is expected to operate in the absence of deviations at the study nodes. This can take a number of forms and can either be descriptive or diagrammatic; e.g., flowsheets, line diagrams, P&IDs.

c. DEVIATIONS- These are departures from the intention which are discovered by systematically applying the guide words (e.g., "more pressure").

d. CAUSES- These are the reasons why deviations might occur. Once a deviation has been shown to have a credible cause, it can be treated as a meaningful deviation. These causes can be hardware failures, human errors, an unanticipated process state (e.g., change of composition), external disruptions (e.g., loss of power), etc.

e. CONSEQUENCES- These are the results of the deviations should they occur (e.g., release of toxic materials). Trivial consequences, relative to the study objective, are dropped.

f. GUIDE WORDS- These are simple words which are used to qualify or quantify the intention in order to guide and stimulate the brainstorming process and so discover deviations. The guide words shown in Table 4-3 are the ones most often used in a HazOp; some organizations have made this list specific to their operations, to guide the team more quickly to the areas where they have previously found problems. Each guide word is applied to the process variables at the point in the plant (study node) which is being examined. For example:

Guide Words		Parameter	Deviation
NO	&	FLOW---------	NO FLOW
MORE	&	PRESSURE-----	HIGH PRESSURE
AS WELL AS	&	ONE PHASE----	TWO PHASE
OTHER THAN	&	OPERATION----	MAINTENANCE

These guide words are applicable to both the more general parameters (e.g., react, transfer) and the more specific parameters (e.g., pressure,

TABLE 4-3. HAZOP GUIDE WORDS AND MEANINGS

Guide Words	Meaning
No	Negation of the Design Intent
Less	Quantitative Decrease
More	Quantitative Increase
Part Of	Qualitative Decrease
As Well As	Qualitative Increase
Reverse	Logical Opposite of the Intent
Other Than	Complete Substitution

temperature). With the general parameters, meaningful deviations are usually generated for each guide word. Moreover, it is not unusual to have more than one deviation from the application of one guide word. For example, "more reaction" could mean either than a reaction takes place at a faster rate, or that a greater quantity of product results.

With the specific parameters, some modification of the guide words may be necessary. In addition, it is not unusual to find that some potential deviations are eliminated by physical limitation. For example, if the design intention of a pressure or temperature is being considered, the guide words "more" or "less" may be the only possibilities.

There are other useful modifications to guide words such as:

- SOONER or LATER for OTHER THAN when considering time
- WHERE ELSE for OTHER THAN when considering position, sources, or destination
- HIGHER and LOWER for MORE and LESS when considering elevations, temperatures, or pressures.

Finally, when dealing with a design intention involving a complex set of interrelated plant parameters (e.g., temperatures, reaction rates, composition, or pressure), it may be better to apply the whole sequence of guide words to each parameter individually than to apply each guide word across all of the parameters as a group. Also, when applying the guide words to a sentence it may be more useful to apply the sequence of guide words to each word or phrase separately, starting with the key part which describes the activity (usually the verbs or adverbs). These parts of the sentence usually are related to some impact on the process parameters. For example, in the sentence "The operator starts flow A when pressure B is reached", the guide words would be applied to:

- flow A (no, more, less, etc.)
- when pressure B is reached (sooner, later, etc.)

Guidelines for Using Procedure

The concepts presented above are put into practice in the following steps:

1. Define the purpose, objectives, and scope of the study
2. Select the team
3. Prepare for the study
4. Carry out the team review
5. Record the results.

It is important to recognize that some of these steps can take place at the same time. For example, the team reviews the design, records the findings, and follows up on the findings continuously. Nonetheless, each step will be discussed below as separate items.

1. Define the Purpose, Objectives, and Scope of the Study. The purpose, objectives, and scope of the study should be made as explicit as possible. These objectives are normally set by the person responsible for the plant or project, assisted by the HazOp study leader (perhaps the plant or corporate safety officer). It is important that this interaction take place to

provide the proper authority to the study and to ensure that the study is focused. Also, even though the general objective is to identify hazards and operability problems, the team should focus on the underlying purpose or reason for the study. Examples of reasons for a study might be to:

- Check the safety of a design
- Decide whether and where to build
- Develop a list of questions to ask a supplier
- Check operating/safety procedures
- Improve the safety of an existing facility
- Verify that safety instrumentation is reacting to best parameters.

It is also important to define what specific consequences are to be considered:

- Employee safety (in plant or neighboring research center)
- Loss of plant or equipment
- Loss of production (lose competitive edge in market)
- Liability
- Insurability
- Public safety
- Environmental impacts.

For example, a HazOp might be conducted to determine where to build a plant to have the minimal impact on public safety. In this case, the HazOp should focus on deviations which result in off-site hazards.

2. Select the Team. Ideally, the team consists of five to seven members, although a smaller team could be sufficient for a smaller plant. If the team is too large, the group approach fails. On the other hand, if the group is too small, it may lack the breadth of knowledge needed to assure completeness. The team leader should have experience in leading a HazOp. The rest of the team should be experts in areas relevant to the plant operation. For example, a team might include:

- Design engineer
- Process engineer
- Operations supervisor
- Instrument design engineer
- Chemist
- Maintenance supervisor
- Safety engineer (if not HazOp leader).

The team leader's most important job is to keep the team focused on the key task: to identify problems, not necessarily to solve them. There is a strong tendency for engineers to launch into a design or problem-solving mode as soon as a new problem comes to light. Unless obvious solutions are apparent, this mode should be avoided or it will detract from the primary purpose of HazOp, which is hazard identification.

In addition, the team leader must keep several factors in mind to assure successful meetings: (1) do not compete with the members; (2) take care to listen to all of the members; (3) during meetings, do not permit anyone to be put on the defensive; (4) to keep the energy level high, take breaks as needed.

3. <u>Prepare for the Study</u>. The amount of preparation depends upon the size and complexity of the plant. The preparative work consists of three stages: obtaining the necessary data; converting the data to a suitable form and planning the study sequence; and arranging the meetings.

a. <u>Obtain the necessary data</u>. Typically, the data consist of various drawings in the form of line diagrams, flowsheets, plant layouts, isometrics, and fabrication drawings. Additionally, there can be operating instructions, instrument sequence control charts, logic diagrams, and computer programs. Occasionally, there are plant manuals and equipment manufacturers' manuals. The data must be inspected to make sure they pertain to the defined area of study and contain no discrepancies or ambiguities.

b. <u>Convert the data into a suitable form and plan the study</u>
    <u>sequence</u>. The amount of work required in this stage depends on
    the type of plant. With continuous plants, the preparative
    work is minimal. The existing, up-to-date flowsheets or pipe
    and instrument drawings usually contain enough information for
    the study, and the only preparation necessary is to make sure
    that enough copies of each drawing are available. Likewise,
    the sequence for the study is straightforward. The study team
    starts at the beginning of the process and progressively works
    downstream, applying the guide words at specific study nodes.
    These nodes are established by the team leader prior to any
    meetings. The team leader will generally define the study nodes
    in pipe sections. These nodes are points where the process
    parameters (pressure, temperature, flow, etc.) have an identi-
    fied design intent. Between these nodes are found the plant
    components (pumps, vessels, heat exchangers, etc.) that cause
    changes in the parameters between nodes. While the study nodes
    should be identified before the meetings, it is to be expected
    that some changes will be made as the study progresses due to
    the learning process that accompanies the study.

    With batch plants, the preparative work is usually more exten-
    sive, primarily because of the more extensive need for manual
    operations; thus, operation sequences are a larger part of the
    HazOp. This operations information can be obtained from opera-
    ting instructions, logic diagrams, or instrument sequence dia-
    grams. In some circumstances (e.g., when two or more batches
    of material are being processed at the same time), it may be
    necessary to prepare a display indicating the status of each
    vessel on a time basis. If operators are physically involved
    in the process (e.g., in charging vessels) rather than simply
    controlling the process, their activities should be represented
    by means of process flow charts.

The team leader will usually prepare a plan for the sequence of study before the study starts to make sure that the study team approaches the plant and its operation methodically. This means the team leader must spend some time before the meetings to determine the "best" study sequence, based on how the specific plant is operated.

The team leader will often have to prepare a representation of the equipment (logic diagram, flow chart, etc.) tailored to suit the application of the HazOp technique to the equipment. This may include a display of the relationship of the equipment with operators and with other plant equipment. The preparative work will often involve a lengthy dialogue between the project engineer and the team leader and sometimes involves the component manufacturers as well. The team leader will prepare a plan for the study and discuss the equipment representations and the plan with the team before starting the study.

c. Arrange the necessary meetings. Once the data have been assembled and the equipment representations made (if necessary), the team leader is in a position to plan meetings. The first requirement is to estimate the team-hours needed for the study. As a general rule, each individual part to be studied, e.g., each main pipeline into a vessel, will take an average of fifteen minutes of team time. For example, a vessel with two inlets, two exits, and a vent should take one and a half hours for those elements and the vessel itself. Thus, an estimate can be made by considering the number of pipelines and vessels. Another way to make a rough estimate is to allow about three hours for each major piece of equipment. Fifteen minutes should also be allowed for each simple verbal statement such as "switch on pump", "motor starts", or "pump starts".

After estimating the team-hours required, the team leader can arrange meetings. Ideally, each session should last no more than three hours (preferably in the morning). Longer sessions are undesirable because their effectiveness usually begins to fall off. Under extreme time-pressures, sessions have been held for two consecutive days; but such a program should be attempted only in very exceptional circumstances, (for example, when the team is from out of town and travel every day is not acceptable.)

With large projects, it has been found that often one team cannot carry out all the studies within the allotted time. It may therefore be necessary to use several teams and team leaders. One of the team leaders should act as a coordinator to allocate sections of the design to different teams and to prepare time schedules for the study as a whole.

4. Carry Out the Team Review. The HazOp study requires that the plant schematic be divided into study nodes and that the process at these points be addressed with the guide words. As shown in Figure 4-4, the method applies all of the guide words in turn and either of two outcomes is recorded: (1) more information is needed, or (2) the deviation with its causes and consequences. If there are obvious remedies, these too are recorded. As hazards are detected, the team leader should make sure that everyone understands them. As mentioned earlier, the degree of problem-solving during the examination sessions can vary. There are two extreme positions:

- A suggested action is found for each hazard as it is detected before looking for the next hazard
- No search for suggested actions is started until all hazards have been detected.

In practice, there is a compromise. It may not be appropriate or even possible for a team to find a solution during a meeting. On the other hand, if the solution is straightforward, a decision can be made and the design and operating instructions modified immediately. To some extent, the ability to make immediate decisions depends upon the type of plant being studied. With a continuous plant, a decision made at one point in a design may not invalidate previous decisions concerning upstream parts of the plant which have already

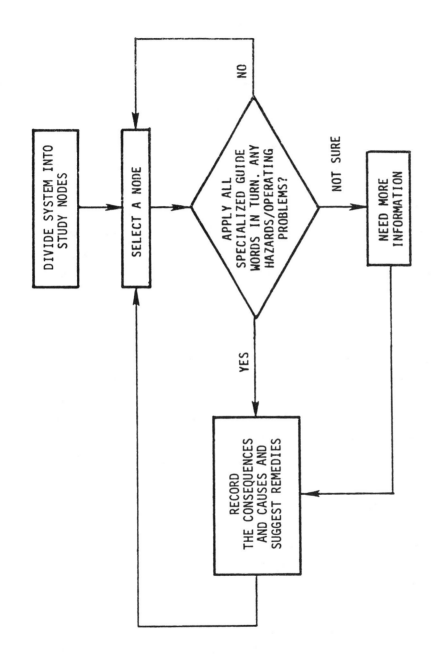

FIGURE 4-4. HAZOP METHOD FLOW DIAGRAM

been studied--but this possibility always has to be considered. For batch plants with sequence control, any alteration in the design or mode of operation could have extensive implications. If a question is noted for future evaluation, a note is also made of the person responsible for follow-up.

Although the team leader will have prepared for the study, the HazOp technique may expose gaps in the available plant operating information or in the knowledge of the team members. Thus it may sometimes be necessary to call in a specialist on some aspects of how the plant is intended to operate or even to postpone certain parts of the study in order to obtain more information.

Once a section of pipeline or a vessel or an operating instruction has been fully examined, the team leader should mark (e.g., "yellow out") his or her copy to that effect. This action ensures comprehensive coverage. Another way of doing this is that once every part of a drawing has been examined, the study leader certifies that the examination has been completed in an appropriate box on the flowsheet.

5. <u>Record the Results</u>. The recording process is an important part of the HazOp. It is impossible to record manually all that is said, yet it is very important that all ideas are kept. It is very useful to have the team members review the final report and then come together for a report review meeting. The process of reviewing key findings will often fine-tune these findings and uncover others. The success of this process demands a good recording scheme.

First, a HazOp form should be filled out during the meeting (a sample is given in Figure 4-5). This form is best filled out by an engineer who can be less senior than the team members. This recorder is not necessarily part of the team but, as an engineer, can understand the discussions and record the findings accurately. Other means of recording can be developed as best suits the organization. Some have found that when insufficient information is available to make a decision, cards are filled out so that the responsible individual is reminded of the action item. It has also been found useful to tape-record the sessions and have them transcribed. This saves the

Process Unit: DAP Production

Node: 1 Process Parameter: Flow

GUIDE WORD	DEVIATION	CONSEQUENCES	CAUSES	SUGGESTED ACTION
No	No Flow	Excess ammonia in reactor. Release to work area.	(1) Valve A fails closed	Automatic closure of valve B on loss of flow from phosphoric acid supply
			(2) Phosphoric acid supply exhausted	
			(3) Plug in pipe; pipe ruptures	
Less	Less Flow	Excess ammonia in reactor. Release to work area, with amount released related to quantitative reduction in supply. Team member to calculate toxicity vs. flow reduction.	(1) Valve A partially closed	Automatic closure of valve B on reduced flow from phosphoric acid supply. Set point determined by toxicity vs. flow calculation
			(2) Partial plug or leak in pipe	
More	More Flow	Excess phosphoric acid degrades product. No hazard to work area.	--	--
Part of	Normal flow of decreased concentration of phosphoric acid	Excess ammonia in reactor. Release to work area, with amount released related to quantitative reduction in supply.	(1) Vendor delivers wrong material or concentration	Check phosphoric acid supply tank concentration after charging
			(2) Error in charging phosphoric acid supply tank	

FIGURE 4-5. SAMPLE OF HAZOP WORKSHEET

only complete record of the discussions and the reasoning behind the recorded findings, and it can be invaluable later in the plant life when the plant is modified, or if an event occurs which is the result of a deviation.

Example

Consider, as a simple example, the continuous process shown in Figure 4-6. In this process, the phosphoric acid and ammonia are mixed, and a non-hazardous product, diammonium phosphate (DAP), results if the reaction of ammonia is complete. If too little phosphoric acid is added, the reaction is incomplete, and ammonia is produced. Too little ammonia available to the reactor results in a safe but undesirable product. The HazOp team is assigned to investigate "Personnel Hazards from the Reaction".

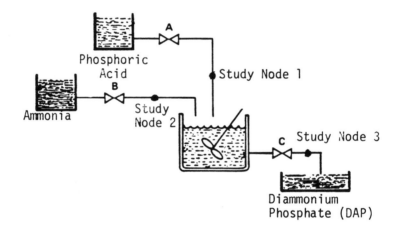

FIGURE 4-6. CONTINUOUS PROCESS EXAMPLE FOR HAZOP TECHNIQUE

The team leader starts with a study node and applies the guide words to the process parameters. Thus, for study node 1:

I. NO

    a.   NO & FLOW --- no flow at study node 1

    b.   Consequences: excess ammonia in reactor and release to work area

    c.   Causes
         -- valve A fails closed
         -- phosphoric acid supply exhausted
         -- plug in pipe, pipe ruptures

    d.   Suggested action: automatic closure of valve B on loss of flow from phosphoric acid supply.

II. LESS

    a.   LESS & FLOW --- reduced flow at study node 1

    b.   Consequences: excess ammonia in reactor and release to work area. Amount released is related to quantitative reduction in supply. Team member assigned to calculate toxicity level versus flow reduction.

    c.   Causes:
         -- valve A partially closed
         -- partial plug or leak in pipe

    d.   Suggested action: automatic closure of valve B based on reduced flow in pipe from phosphoric acid supply. Set point dependent on toxicity versus reduced flow calculations.

III. MORE

    a.   MORE & FLOW --- increased flow at study node 1

    b.   Consequences: excess phosphoric acid degrades product but presents no hazard to workplace.

IV. PART OF

    a.   PART OF & FLOW --- decreased concentration of phosphoric acid at study node 1

    b.   Consequences: see II.b (low flow consequences)

    c.   Causes
         -- vendor delivers wrong material or concentration
         -- error in charging phosphoric acid supply tank

    d.   Suggested Action: Add check of phosphoric acid supply tank concentration after charging procedures.

V. AS WELL AS

    a.   AS WELL AS & FLOW --- increase concentration of
        phosphoric acid
        (not a realistic consideration since highest available
        concentration used to charge supply).

VI. REVERSE

    a.   REVERSE & FLOW --- reverse flow at study node 1
    b.   Consequences
    c.   Causes: no reasonable mechanism for reverse flow.

VII. OTHER THAN

    a.   OTHER THAN & FLOW --- material other than phosphoric
        acid in line A
    b.   Consequences: Depends on substitution; team member
        assigned to test potential substitutions based on avail-
        ability of other materials at site and similarity in
        appearance
    c.   Causes:
        -- wrong delivery from vendor
        -- wrong material chosen from plant warehouse
    d.   Recommended Action: Plant procedures to provide check on
        material chosen before charging phosphoric acid supply
        tank.

This process then continues by choosing other process parameters and combining
them with the guide words.

HazOp Variations

1. Knowledge-Based HazOp

    The knowledge-based HazOp is a specialization of the Guide-Word
HazOp in which the guide words are replaced by the team's and leader's knowledge

and specific checklists. This knowledge base is used to compare the design to established basic design practices which have been developed and documented from previous plant experience. The implicit premise of this version of the HazOp is that the organization has extensive design standards and the team members are familiar with them. An important advantage of this method is that the lessons learned over many years of experience are incorporated into the company's practices and are thus available for use at all stages in the design and construction of the plant. Thus, the Knowledge-Based HazOp study can help ensure that the company's practices, and therefore its past experience, have indeed been incorporated in the design.

Comparison of a design in the Knowledge-Based HazOp with codes and practices will generate a set of questions which are different from the Guide-Word HazOp. For example, questions might be:

- "Shouldn't the design be like ...?"
- "Will this change cover the hazard at the same risk level?"

As a more specific example, consider the discharge from a centrifugal pump. The Guide Word HazOp would apply the guide word REVERSE to identify the need for a check valve. The Knowledge-Based HazOp would identify the need for a check valve because an actual problem was experienced with reverse flow and the use of check valves on a centrifugal pump discharge has been adopted as a standard practice. Figure 4-7 shows an expanded checklist for centrifugal pumps from one company's Knowledge-Based HazOp checklist. The complete checklist can be found in Appendix A.

It is also important to note that the Guide-Word HazOp approach can be used to supplement this approach to ensure that new problems are not overlooked when portions of the process involve major changes in equipment technology or novel chemistry. (See C. H. Solomon for more details.)

1. Can Casing Design Pressure be Exceeded?
   - Maximum suction pressure + shutoff delta P
     - Note: If pump curve not available, shutoff delta P (motor drive) may be estimated as 120 percent of operating delta P and shutoff delta P (turbine-drive) may be estimated as 132 percent of operating delta P.
   - Higher than design specific gravity of pumped fluid
     - During startup
     - During upsets

2. Is Downstream Piping/Equipment Adequately Rated?
   - If downstream blockage raises suction pressure:
     DP = maximum suction P + shutoff delta P
   - If downstream blockage does not raise suction pressure,
     DP = the greater of:
         normal suction plus shutoff delta P or
         maximum suction plus normal delta P

3. Is Backflow Prevented?
   - Check valve in discharge
   - Double check valve for delta P >1,000 psi

4. Suction Piping Overpressure (single pumps)
   - Suction valve, flange, and connecting piping same as suction line DP

5. Suction Piping Overpressure (parallel pumps)
   - Suction valve and intervening component DP
     >3/4 pump discharge DP

6. Is Damage from Low Flow Prevented?
   - Recycle system to ensure 20 percent best-efficiency-point flow

7. Can Fire Be Limited?
   - Provide isolation valve(s) if suction vessel(s) inventory is:
     - light ends > 2,000 gallons, or
     - HC liquid above 600°F > 2,000 gallons, or
     - HC liquid >4,000 gallons
   - Remote actuated if
     - 10" or larger line size, or
     - located in fire risk area (<25 ft. horizontal)
       Note: Valve activator and exposed cable fireproofed

FIGURE 4-7. SAMPLE WORKSHEET FOR KNOWLEDGE-BASED HAZOP: CENTRIFUGAL PUMPS

2. Creative Checklist HazOp

The Creative Checklist HazOp study was developed to address two needs:

- The need for a study that can be carried out earlier in the design, based only on the materials to be used
- The need for a study that can examine adverse interactions resulting from the proximity of the units of the plant, or interaction of the units and the environment.

This form of HazOp utilizes the same basic form as the Guide Word HazOp listed above.

At an early stage in a project, the only data available could be a list of all material likely to be used, together with a site plan and a block layout of the site.

First, the materials are compared to a hazards checklist (fire, toxicity, reactivity) and the team determines which hazards could really exist. The hazardous materials are inventoried, and such information as the likely volume of each material is listed.

The second part of this study is to associate each unit of the site plan with the hazards list created in the first part. This results in a series of hypothetical "unit hazards". If the team feels that a given unit hazard is real, they identify potential actions and/or guidelines which should be followed in the subsequent design phase to minimize the risks. Because the study takes place before the site has been finalized, a decision to abandon a site, if necessary, could be made during this study.

This HazOp is very similar to the Preliminary Hazard Analysis (PHA) described in this document. It does have the advantage of conducting the study of an early design by a team rather than by one or two people. Of course, it will also be more expensive because more people are involved.

References

Solomon, C. H., "The Exxon Chemicals Method for Identifying Potential Process Hazards", I Chem E Loss Prevention Bulletin No. 52, August 1983.

Knowlton, E., "Creative Checklist Hazard and Operability Studies," Chemical Manufacturers Association, Process Safety Management Workshop, Arlington, Virginia, May 1985.

4.7 Failure Modes, Effects, and Criticality Analysis

Description

Failure Modes, Effects, and Criticality Analysis (FMECA) is concerned almost entirely with equipment, the ways in which it could fail, the effects that could be generated, and the estimated failure probabilities. These failure data provide the basis for determining where changes can be made most advantageously to improve the probability that a design will function successfully. Hazards and the possibilities of damage are related to failures only; they rarely involve investigation into damage or injury that could arise if the system operated successfully.

Each failure is considered individually as an independent occurrence, with no relation to other failures in the system except the subsequent effects that it might produce. Generally, FMECA is first accomplished on a qualitative basis. Quantitative data may then be applied to establish a criticality ranking for the system or subsystem.

Criticality rankings are generally expressed as probabilities but may also be indicated in other ways. In some instances, they are designated in categories from 1 to 10 (or from A to Z) to show the principal items that could generate problems. These categories are often not based on probabilities but reflect experience.

Guidelines for Using Procedure

The FMECA procedure contains five steps:

1. Determine level of resolution
2. Develop a consistent format
3. Define the problem and boundary conditions
4. Complete the FMECA table
5. Report the results.

Each of these is discussed below.

1. Determine Level of Resolution

The level of resolution determines the detail to be included in the FMECA tables. If a plant-level hazard is being addressed, the FMECA should focus on the individual systems or subsystems in the plant and on their failure modes and effects with respect to the plant-level hazard; for example, the FMECA might focus on the feed system, batch mixing system, oxidizing system, product separation system, and the various support systems that make up the plant. When a system-level hazard is being addressed, the FMECA should focus on individual equipment that makes up the system and on its failure modes and effects with respect to the system-level hazard. For a system-level hazard, such as loss of temperature control in the oxidizing system, the FMECA might focus on the feed pump, cooling water pump, cooling water flow control valve, and temperature sensor and alarm that make up the oxidizing system. Of course, effects identified at the system or equipment level may subsequently be related to potential plant hazards in the FMECA tables.

2. Develop a Consistent Format

A standard FMECA format promotes consistency in the information contained in the FMECA tables and assists in maintaining the level of resolution defined in Step 1. Figure 4-8 shows an example format for an FMECA table. Additional information, such as the failure mode probability, may be included in the tables to support the criticality ranking definition or other types of hazard assessment. For example, equipment failure probability may be entered in the table to provide a reference source for subsequent quantitative analyses.

Item	Identification	Description	Failure Modes	Effects	Criticality Ranking

DATE: _____

PLANT: _____

SYSTEM: _____

PAGE _____ OF _____

REFERENCE: _____

FIGURE 4-8. SAMPLE FORMAT FOR AN FMECA TABLE

3. Define the Problem and Boundary Conditions

This step identifies the specific items to be included in the FMECA within the previously defined level of resolution. The problem and boundary condition definition specifically states what systems and equipment are to be included in the FMECA. Minimum requirements for the problem definition include:

- Identifying the plant and/or systems that are the subject of the analysis.
- Establishing the physical system boundaries that encompass the equipment contained in the FMECA. This statement specifies the interfaces with other processes and utility/support systems and what portions of these interfaces are to be included in the FMECA. One way to indicate the physical system boundaries is to mark them on a system drawing that encompasses all equipment in the FMECA. These boundary conditions should also state the operating conditions at the interface that are assumed for the FMECA.
- Collecting up-to-date reference information that identifies the equipment and its functional relationship to the plant/system. This information is needed for all equipment included within the system boundary.
- Providing a consistent criticality ranking definition that addresses the potential effects of the equipment failures. Table 4-4 provides an example of a criticality ranking definition. The criticality ranking may be defined in terms of the probability of the failure, the severity of the resulting accident, or a combination of these factors. The problem definition may also include other facility- or process-specific assumptions that have a direct influence on the effects resulting from equipment failures.

TABLE 4-4. EXAMPLE OF CRITICALITY RANKING DEFINITIONS FOR FMECA

Effects	Criticality Ranking
None	1 (best)
Minor process upset, small hazard to facilities and personnel, process shutdown not required	2
Major process upset, signifi- cant hazard to facilities and personnel, orderly process shutdown required	3
Immediate hazard to facilities and personnel, emergency shutdown required	4 (worst)

4. Complete the FMECA Table

The FMECA table should be completed in a deliberate, systematic manner to reduce the possibility of omissions and to enhance the completeness of the FMECA. A table can be produced by beginning at a system boundary on a reference drawing and systematically evaluating the items in order as they appear in the process flowpath. Each equipment item can then be checked off or "red-lined" on the reference drawing when its failure modes have been evaluated completely. All entries for each item or system being addressed in the FMECA should be completed before proceeding to the next item. The following items should be standard entries in the FMECA table:

Equipment Identification: A unique equipment identifier that relates the equipment to a system drawing, process, or location. This identifier distinguishes between similar equipment (e.g., two motor-operated valves) that perform different functions within the same system. Equipment numbers or identifiers from system drawings, such as piping and instrumentation diagrams, are usually available and provide a reference to existing system information. Practically any systematic coding scheme is acceptable as long as the identifiers are meaningful to the analysts who must work with the results of the FMECA.

Equipment Description: The equipment description should include the equipment type, operating configuration, and other service characteristics (such as high temperature, high pressure, or corrosive service) that may influence the failure modes and their effects: for example, "motor-operated valve, normally open, three-inch sulfuric acid line". These descriptions need not be unique for each item of equipment.

Failure Modes: The analyst should list all failure modes for each item consistent with the equipment description. Considering the equipment's normal operating condition, the analyst should consider all conceivable malfunctions that alter the equipment's normal operation. For example, the failure modes of a normally open valve may include:

• Fails open (or fails to close when required)
• Transfers to a closed position
• Leaks to external environment
• Valve body ruptures.

Table 4-5 contains additional examples of equipment failure modes. The analyst should concentrate on identifying the various failure modes rather than the potential causes of the failure. Considering various causes will assist in identifying different failure modes. However, the analyst should limit the table entries to failure modes even though there may be several causes of the failure mode. The analyst should include all postulated failure modes so that their effects can be addressed.

TABLE 4-5. EXAMPLE OF EQUIPMENT FAILURE MODES FOR FMECA

Equipment Description	Failure Modes
Pump, normally operating	a. Fails on (fails to stop when required) b. Transfers off c. Seal rupture/leak d. Pump casing rupture/leak
Heat exchanger, high pressure on tube side	a. Leak/rupture, tube side to shell side b. Leak/rupture, shell side to external environment c. Tube side, plugged d. Shell side, plugged

Effects: For each identified failure mode, the analyst should describe both the immediate and expected effects of the failure on other equipment and the process or system. For example, the immediate effect of a pump seal leak is a spill in the area of the pump. If the fluid is flammable, a fire could be expected (because the pump is a potential ignition source) that might involve additional nearby equipment. The analyst may also note the expected response of any applicable safety systems that could mitigate the effect. For example, the effects associated with a cooling water valve transferring to a closed position might be stated as: "Loss of cooling water flow to process vessel #1, resulting in overheating of the process fluid. Vessel #1 high-temperature alarm setpoint is 350 F".

Criticality Ranking: The analyst should classify each failure mode and effect according to the criticality ranking definition developed in the problem definition. Each effect is examined in terms of its hazard and the potential result of that hazard and then compared to the ranking definition for classification.

5. Report the Results

The result of the FMECA is a systematic and consistent tabulation of the effects of equipment failure within a process or system. The equipment identification in the FMECA provides a direct reference between the equipment and system piping and instrumentation drawings or process flow diagrams. The criticality ranking provides a relative measure of the equipment failure mode's contribution to the system hazards.

Equipment failures with an unacceptable criticality ranking should be re-examined to verify the failure modes and their effects. These failures are the most likely candidates for protective measures, especially if the failure leads directly to a serious accident.

The results of the FMECA are useful in other hazard evaluation methods. For example, in conjunction with a HazOp study, the FMECA provides a concise summary of the hazards associated with component failures. (In fact, the FMECA is a subset of a complete HazOp study.) The FMECA is also useful in fault tree analysis, event tree analysis, and cause-consequence analysis, where the analyst must determine the contributing equipment failure for a stated hazard. For example, an important hazard identified in a HazOp can be compared to the effects listed in the FMECA to identify specific equipment failure modes that are directly involved with the hazard.

References

1. Hammer, W., Handbook of System and Product Safety, Prentice-Hall, Inc., 1972.

2. Lambert, H. E., Failure Modes and Effects Analysis, NATO Advanced Study Institute, 1978.

3. Lees, F. P., Loss Prevention in the Process Industries, London: Butterworths, 1980.

4. McCormick, N. J., Reliability and Risk Analysis, New York: Academic Press, Inc., 1981.

4.8 Fault Tree Analysis

General Description

Fault Tree Analysis (FTA) is a widely used tool for system safety analysis. One of the primary strengths of the method is the systematic, logical development of the many contributing failures that might result in an accident. This type of development requires that the analyst have a complete understanding of the system/plant operation and the various equipment failure modes.

FTA breaks down an accident into its contributing equipment failures and human errors. The method therefore is a "reverse-thinking" technique; that is, the analyst begins with an accident or undesirable event that is to be avoided and identifies the immediate causes of that event. Each of the immediate causes is examined in turn until the analyst has identified the basic causes of each event. The fault tree is a diagram that displays the logical interrelationships between these basic causes and the accident.

The result of the FTA is a list of combinations of equipment and human failures that are sufficient to result in the accident. These combinations of failures are known as minimal cut sets. Each minimal cut set is the smallest set of equipment and human failures that are sufficient to cause the accident if all the failures in that minimal cut set exist simultaneously. Thus, a minimal cut set is logically equivalent to the undesired accident stated in terms of equipment failures and human errors.

Logic Symbols for Fault Tree Analysis. The fault tree is a graphical representation of the interrelationships between equipment failures and a specific accident. The following symbols are used in fault tree construction to display these relationships.

OR Gate: The OR gate indicates that the output event occurs if any of the input events occur.

AND Gate: The AND gate indicates that the output event occurs only when all the input events occur.

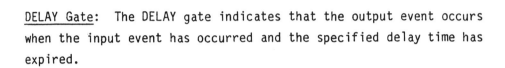 INHIBIT Gate: The INHIBIT gate indicates that the output event occurs when the input event occurs and the inhibit condition is satisifed.

 DELAY Gate: The DELAY gate indicates that the output event occurs when the input event has occurred and the specified delay time has expired.

 BASIC Event: The BASIC event represents a basic equipment fault or failure that requires no further development into more basic faults or failures.

 INTERMEDIATE Event: The INTERMEDIATE event represents a fault event that results from the interactions of other fault events that are developed through logic gates such as those defined above.

UNDEVELOPED Event: The UNDEVELOPED event represents a fault event that is not examined further because information is unavailable or because its consequence is insignificant.

EXTERNAL or HOUSE Event: The EXTERNAL or HOUSE event represents a condition or an event that is assumed to exist as a boundary condition for the fault tree.

 TRANSFER Symbol: The TRANSFER IN symbol indicates that the fault tree is developed further at the occurrence of the corresponding TRANSFER OUT symbol (e.g., on another page). The symbols are labelled using numbers or a code system to ensure that they can be differentiated.

Definitions. Equipment faults and failures that are described in a fault tree can be grouped into three classes:

1. primary faults and failures
2. secondary faults and failures
3. command faults and failures.

Primary faults and failures are equipment malfunctions that occur in the environment for which the equipment was intended; for example, a pressure vessel ruptures at a pressure that is within the design limits of the vessel. Primary faults and failures are the responsibility of the equipment that failed and cannot be attributed to some external force or condition.

Secondary faults and failures are equipment malfunctions that occur in an environment for which the equipment was not intended; for example, a pressure vessel ruptures because some fault external to the vessel causes the internal pressure to exceed the design limits of the vessel. Secondary faults and failures are not the responsibility of the equipment that failed but can be attributed to some external force or condition.

Command faults and failures are equipment malfunctions in which the component operates properly but at the wrong time or in the wrong place; for example, a temperature alarm fails to announce high temperature in a process because the temperature sensor has failed. The alarm failure is a command failure; that is, the failed sensor (a primary failure) commands the alarm not to sound when high temperature occurs. Command faults and failures are not the responsibility of the equipment that is commanded to fail, but can be attributed to the source of the incorrect command.

All three classes of faults and failures will normally appear in a fault tree. One of the objectives of fault tree analysis is to identify the basic contributing failures that result in an accident. These basic contributing failures are primary faults and failures identifying the equipment that is responsible for the failure. Secondary and command faults and failures are the intermediate events in the fault tree that require additional development to identify the primary faults and failures (BASIC events) that result in the intermediate event.

Guidelines for Using Procedure

There are four steps in performing a fault tree analysis: problem definition, fault tree construction, fault tree solution (determining minimal cut sets), and minimal cut set ranking.

1. Problem Definition

The problem definition consists of:

* Defining a TOP Event, the accident event that is the subject of the fault tree analysis
* Defining analysis boundary conditions, including
    - unallowed events
    - existing events
    - system physical bounds
    - level of resolution
    - other assumptions.

Defining the TOP event is one of the most important aspects of the problem definition. The TOP event is the accident or undesired event that is the subject of the fault tree analysis. This event should be precisely defined for the system or plant under study. Vague or poorly defined TOP events often lead to a diverging analysis. For example, a TOP event of "Fire at the Plant" is too general to define a fault tree analysis. The TOP event should be something like "Fire in the Process Oxidation Reactor During Normal Operation". This event description gives the three necessary items in the event description: what, where, and when. The what (fire) tells us the type of accident, the where (process oxidation reactor) tells us the system or process equipment involved in the accident, and the when (during normal operation) tells us the overall system configuration.

The analysis boundary conditions are necessary to define the events that will eventually make up the fault tree. These boundary conditions describe the system in its normal, unfailed state. Unallowed events are events that are considered not to be possible for the purposes of this fault tree analysis. Existing events are events that are considered certain to occur for the purposes of this fault tree analysis. Often the unallowed and existing events do not appear in the finished fault tree, but their effects must be considered in developing other fault events as the fault tree is constructed.

Another analysis boundary condition is the description of the system's physical boundaries. The physical system boundaries encompass the equipment that will be considered in the fault tree, and the equipment's interfaces with other processes and utility/support systems that will be included. One way to indicate the physical system boundary is to mark it on the system drawing that encompasses all the equipment in the fault tree analysis. As the fault tree is constructed, equipment can be checked off as its failure development is entered into the fault tree. Along with the physical system boundaries, the analyst should specify the level of resolution for the fault tree events. The level of resolution simply states the amount of detail to be included in the fault tree. For example, a motor-operated valve can be included as a single piece of equipment, or it can be described as several hardware items (valve body, valve internals, and motor operator). This breakdown could also include the necessary switchgear, power supply, and human operator needed for the valve to perform. One factor that should be considered in deciding on the level of resolution is the amount of detail in the failure information that is available to the analyst, perhaps from an FMECA or previous safety study. The resolution of the fault tree should parallel the resolution of the available information.

Another system boundary condition is the initial equipment configuration or initial conditions. This information states the configuration of the system and the equipment that is assumed for the fault tree analysis. The analyst specifies which valves are open, which valves are closed, which pumps are on, which pumps are off, etc., for all equipment within the physical system boundaries.

The analyst may specify other assumptions as necessary to define the system for the fault tree analysis. For example, the analysis may assume that the system is operating at 50 percent of normal capacity. After the completion of the problem definition and all boundary conditions, these assumptions should clear up any questions that remain about the state of the system in the fault tree analysis.

2. Fault Tree Construction

Fault tree construction begins at the TOP event and proceeds, level by level, until all fault events have been developed to their basic contributing causes (BASIC events). The analyst begins with the TOP event and, for the next level, determines the immediate, necessary, and sufficient causes that result in the TOP event. Normally, these are not basic causes but are intermediate faults that require additional development. If the analyst can immediately determine the basic causes of the TOP event, the problem is not suitable (too simple) for fault tree analysis and can be evaluated by other methods such as FMECA. Figure 4-9 shows a representative fault tree structure using the symbols defined above. The immediate causes of the TOP event are shown in the fault tree with their relationship to the TOP event. If any one of the immediate causes results directly in the TOP event, the causes are connected to the TOP event with an OR logic gate. If all the immediate causes are required for TOP event occurrence, then the causes are connected to the TOP event with an AND logic gate. Each of the immediate causes is then treated in the same manner as the TOP event, and its immediate, necessary, and sufficient causes are determined and shown on the fault tree with the appropriate logic gate. This development continues until all intermediate fault events have been developed into their basic causes.

Rules for Fault Tree Construction. Several basic rules for fault tree construction have evolved in the last 15 years to promote consistency and completeness in the fault tree construction process.

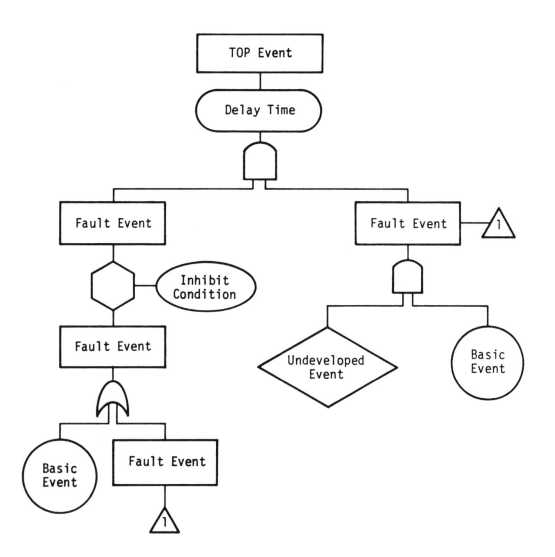

FIGURE 4-9.   REPRESENTATIVE FAULT TREE STRUCTURE

a. <u>Fault Event Statements</u>: Write the statements that are entered in the event boxes and circles as faults; state precisely what the fault is, where the fault is, and when it occurs. Making these statements as precise as possible is necessary for complete description of the fault event. The "where" and "what" portions specify the equipment and its relevant failed state. The "when" condition describes the state of the system with respect to the equipment, thus telling why the equipment state is a fault. These statements must be as complete as possible; the analyst should resist the temptation to abbreviate them during the fault tree construction process.

b. <u>Fault Event Evaluation</u>: When evaluating a fault event, ask the question "Can this fault consist of an equipment failure?". If the answer is "yes", classify the fault event as a "state-of-equipment fault". If the answer is "no", classify the fault event as a "state-of-system fault". This classification aids in the continued development of the fault event. If the event is a state-of-equipment fault, add an OR gate to the fault event and look for primary, secondary, and command failures that can result in the event. If the fault event is a state-of-system fault, then look for the immediate, necessary, and sufficient causes of the fault event.

c. The "No Miracles" Rule: If the normal functioning of equipment propagates a fault sequence, then assume that the equipment functions normally. Never assume that the miraculous and totally unexpected failure of some equipment interrupts or prevents an accident from occurring.

d. The "Complete the Gate" Rule: All inputs to a particular gate should be completely defined before further analysis of any gate. The fault tree should be completed in levels, and each level should be completed before beginning the next level.

e. <u>The "No Gate-to-Gate" Rule</u>: Gate inputs should be properly defined fault events, and gates should not be directly connected to other gates. Shortcutting the fault tree development leads to confusion because the outputs of the individual gates are not specified.

Rules (c) and (e) are intended to emphasize how important it is to be systematic and methodical in constructing fault trees. Shortcuts prohibited by these rules often lead to an incomplete fault tree that over-looks potentially important combinations of equipment failures. These shortcuts also limit the use of the fault tree as a communication tool, because only the analyst who developed the fault tree will be able to decipher the thought process that resulted in its development.

3. Fault Tree Solution

The completed fault tree provides much useful information by displaying the interactions of equipment failures that could result in an accident. However, except for the simplest fault trees, even an experienced analyst cannot identify directly from the fault tree all the combinations of equipment failures that can lead to the accident. This section discusses a method of "solving" the fault tree, or obtaining the minimal cut sets for the fault tree. The minimal cut sets are all the combinations of equipment failures that can result in the fault tree TOP event, and they are logically equivalent to the information displayed in the fault tree. The minimal cut sets are useful for ranking the ways in which the accident may occur, and they allow quantification of the fault tree if appropriate data are available. Large fault trees require computer programs to determine their minimal cut sets; however, the method described here will allow hand-solution of many of the fault trees encountered in practice.

The fault tree solution method has four steps:

a. Uniquely identify all gates and BASIC events
b. Resolve all gates into BASIC events

   c. Remove duplicate events within sets

   d. Delete all supersets (sets that contain another set as a
      subset).

The result of the procedure is the list of minimal cut sets for the fault
tree.

      The procedure is demonstrated with an example. Figure 4-10 is an
example fault tree. Step (a) is to uniquely identify all gates and BASIC
events in the fault tree. In Figure 4-10, the gates are identified with
letters and the BASIC events with numbers. Each identification must be
unique, and if a BASIC event appears more than once in the fault tree, it must
have the same identifier each time. For example, BASIC event 2 appears twice
in Figure 4-10, each time with the same identifier.

      The second step is to resolve all the gates into BASIC events. This
is done in a matrix format, beginning with the TOP event and proceeding
through the matrix until all gates are resolved. Gates are resolved by
replacing them in the matrix with their inputs. The TOP event is always the
first entry in the matrix and is entered in the first column of the first row
(see Figure 4-11a). There are two rules for entering the remaining information in the matrix: the OR-gate rule, and the AND-gate rule.

      <u>The OR-gate rule</u>: When resolving an OR gate in the matrix, the
      first input of the OR gate replaces the gate identifer in the
      matrix, and all other inputs to the OR gate are inserted in the next
      available row, one input per row. The next available row means the
      next empty row of the matrix. In addition, if there are other
      entries in the row where the OR gate appeared, these entries must be
      entered (repeated) in all the rows that contain the gate's inputs.

      <u>The AND-gate rule</u>: When resolving an AND gate in the matrix, the
      first input to the AND gate replaces the gate identifer in the
      matrix, and the other inputs to the AND gate are inserted in the
      next available column, one input per column, on the same row that

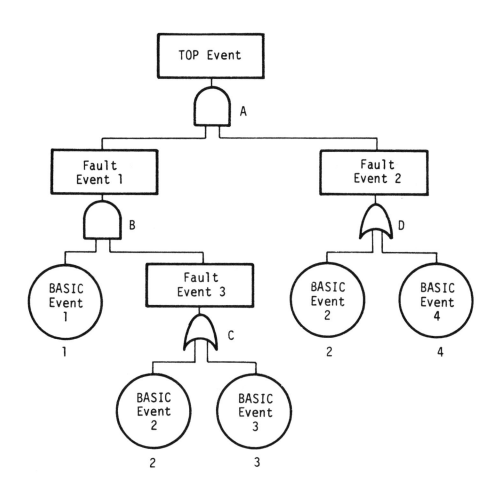

FIGURE 4-10. SAMPLE FAULT TREE

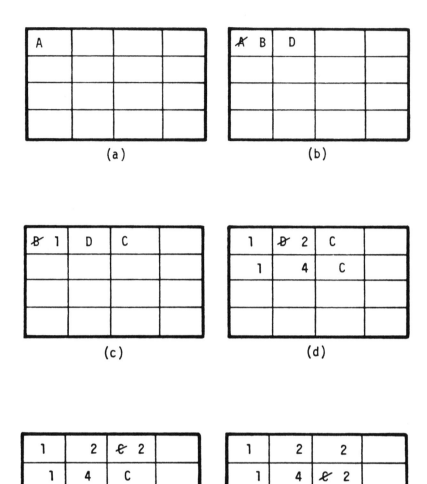

FIGURE 4-11. MATRIX FOR RESOLVING GATES IN SAMPLE FAULT TREE

the AND gate appeared on. INHIBIT and DELAY gates are resolved as
if they were AND gates.

These two rules are repeated as necessary until only BASIC event identifers
remain in the matrix.

Now we will solve the fault tree in Figure 4-10 using these two
rules. Figure 4-11a shows the first entry in the matrix, gate A, which is the
TOP event in our sample fault tree. Gate A is an AND gate, so we apply the
AND-gate rule to resolve gate A into its inputs, gates B and D, as shown in
Figure 4-11b. Now we choose the next gate to resolve, for example, gate B.
Gate B is also an AND gate, so its inputs are entered on the same row as
gate B. This replacement is shown in Figure 4-11c. Next, we resolve gate D.
Gate D is an OR gate, so its first input replaces D and its second input is
entered in the next available row, as shown in Figure 4-11d. Notice that the
other components of row 1 (where gate D appeared) are also entered (repeated)
on the next available row. Gate C is now the only gate left in the matrix,
appearing on both row 1 and row 2. Each occurrence of gate C is resolved
separately. First, on row 1, we apply the OR-gate rule to gate C as shown in
Figure 4-11e, resulting in a new set of entries in row 3. Similarly, we
resolve the second occurrence of gate C as shown in Figure 4-11f. This
completes the resolution of the gates in the matrix. The results of this step
are four sets of BASIC events:

> Set 1 : 1,2,2
> Set 2 : 1,2,4
> Set 3 : 1,2,3
> Set 4 : 1,3,4

These results are the input to the third step of the fault tree
solution procedure, which is to remove duplicate events within each set of
BASIC events identified in step two. Only Set 1 above has a repeated BASIC
event in the results: BASIC event 2 appears twice. When we remove this
repeated event, the sets of BASIC events are:

> Set 1 : 1,2
> Set 2 : 1,2,4
> Set 3 : 1,2,3
> Set 4 : 1,3,4

The fourth step of the fault tree solution procedure is to delete all supersets that appear in the sets of basic events. In the results above there are
two supersets. Both Set 2 and Set 3 are supersets of Set 1; that is, Sets 2
and 3 each contain Set 1 as a subset. Once these supersets are deleted, the
remaining sets are the minimal cut sets for our example fault tree:

> Minimal Cut Set 1 : 1,2
> Minimal Cut Set 2 : 1,3,4

4. Minimal Cut Set Ranking

Ranking the minimal cut sets is the final step of the fault tree
analysis procedure. For a qualitative ranking, two factors can be considered.
First is structural importance, which is based on the number of BASIC events
that are in each minimal cut set. In this type of ranking, a one-event
minimal cut set is more important than a two-event minimal cut set; a two-
event set is more important than a three-event set; and so on. This ranking
implies that one event is more likely to occur than two events, two events are
more likely to occur than three events, etc.

The second factor considers ranking within each size of minimal cut
set; for example, ranking the two-event minimal cut sets, based on what type
of events make up the minimal cut set. The general rule that guides this
ranking is:

1. human error
2. active equipment failure
3. passive equipment failure.

This ranking implies that human errors are more likely to occur than active
equipment failures (functioning equipment, such as a running pump) and that
active equipment failures are more likely to occur than passive equipment
failures (static, non-functioning equipment, such as a storage tank). Using
this rule on a list of two-event minimal cut sets might result in the following ranking:

Rank	Basic Event 1 Type	Basic Event 2 Type
1	human error	human error
2	human error	active equipment failure
3	human error	passive equipment failure
4	active equipment failure	active equipment failure
5	active equipment failure	passive equipment failure
6	passive equipment failure	passive equipment failure

These rankings, although suggested by experience, may differ significantly from system to system based on such factors as the quality of equipment, the inservice test and maintenance policy, and the degree of operator training. The best qualitative ranking method is for an experienced analyst or engineer to examine the individual minimal cut sets and establish the most important sets based on actual operating experience.

Example of Fault Tree Construction

Figure 4-12 shows the system that is analyzed in this example. The system consists of a process reactor for a highly unstable process that is sensitive to a small increase in temperature. The system is equipped with a quench tank to protect the system against an uncontrolled reaction. To prevent damage to the system, the quench tank must flood the reactor or the reactor inlet flow must be stopped. The reactor process is monitored by two temperature sensors, T1 and T2. Sensor T1 automatically activates the quench tank outlet valve when it detects a rise in temperature. Sensor T2, at the same time, sounds an alarm in the control room to alert the operator to the uncontrolled reaction. When the alarm sounds, the operator pushes the inlet valve close button to close the inlet valve and stop the reactor inlet flow. The operator also pushes the quench tank valve button in the control room in case sensor T1 failed to open the valve. If the quench tank valve opens or the inlet valve closes, the reactor enters a stable shutdown with no damage to the system.

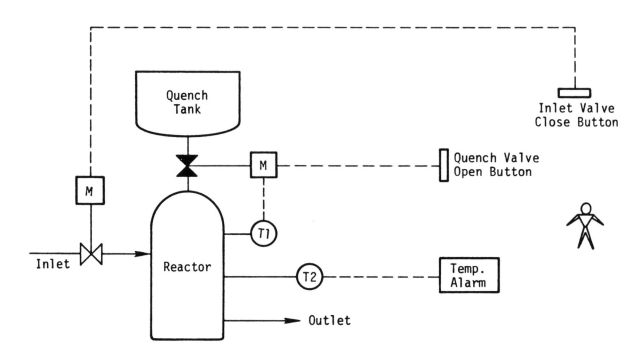

FIGURE 4-12. SYSTEM USED FOR FAULT TREE EXAMPLE

The first step in the fault tree analysis is to define the problem. For this example, the problem definition is as follows.

1. TOP Event: Damage to reactor due to high process temperature.

2. Existing Event: High process temperature.

3. Unallowed Events: Electric power failures; Push button failures; Wiring failures.

4. Physical Bounds: As shown in Figure 4-12. Process upstream or downstream of the reactor is not considered.

5. Equipment Configurations: Inlet valve open, quench tank valve closed.

6. Level of Resolution: Equipment as shown in Figure 4-12.

This problem definition completely describes the system and conditions to be developed in the fault tree.

Fault tree construction begins with the TOP event and proceeds level by level until all faults have been developed to their basic causes. To begin the fault tree, first determine the immediate, necessary, and sufficient causes of the TOP event and identify the logic gate that defines the relationship of those causes to the TOP event. From the example system description, we can identify two causes for the TOP event:

1. No flow from the quench tank
2. Reactor inlet valve remains open.

Since both of these events must occur to result in the TOP event, the development requires an AND logic gate as shown in Figure 4-13.

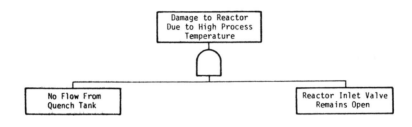

FIGURE 4-13. CAUSES FOR TOP EVENT IN FAULT TREE EXAMPLE

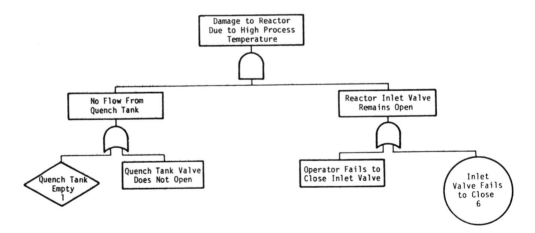

FIGURE 4-14. CAUSES FOR FIRST TWO INTERMEDIATE EVENTS
IN FAULT TREE EXAMPLE

Development now continues to the next level; that is, each of the fault events is developed by determining its immediate, necessary, and sufficient causes. For the event "no flow from quench tank", there are two causes:

1. Quench tank empty
2. Quench tank valve does not open.

Since either of these causes results in no flow, they are added to the fault tree with an OR logic gate. Figure 4-14 shows this development and the continued development of the event "reactor inlet valve remains open."
Figure 4-14 shows the "quench tank empty" event represented by the symbol for an undeveloped event (diamond). There may be several causes for the event, but they are outside our problem definition and are not developed in this fault tree.

Fault tree development continues until all events are developed to basic causes. The completed fault tree is shown in Figure 4-15. As an exercise, the reader is encouraged to continue the development of Figure 4-14 to identify the thought process that results in the completed fault tree.

The next step in fault tree analysis is to determine the minimal cut sets for the fault tree by following the procedure described earlier.
Table 4-6 contains the list of minimal cut sets to allow the reader to check the solution.

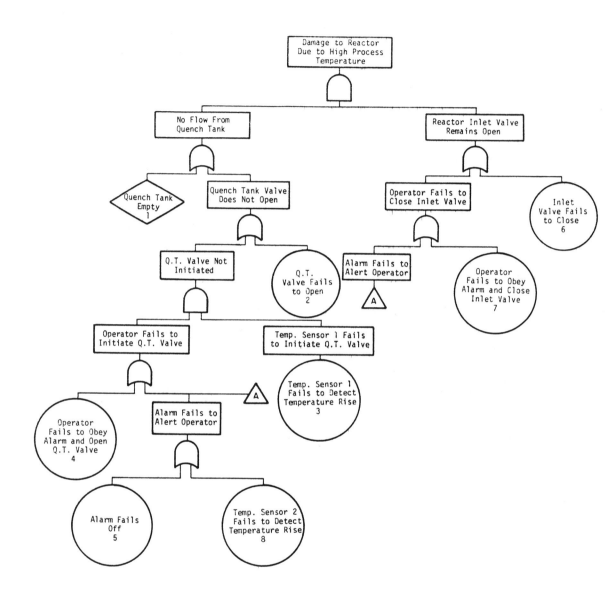

FIGURE 4-15. COMPLETED FAULT TREE

TABLE 4-6. MINIMAL CUT SETS FOR THE
EXAMPLE FAULT TREE

Set No.	Minimal Cut Set
1	1,5
2	1,6
3	1,7
4	1,8
5	2,5
6	2,6
7	2,7
8	2,8
9	3,5
10	3,8
11	3,4,6
12	3,4,7

References

1. Fussell, J. B., "Fault Tree Analysis: Concepts and Techniques", Generic Techniques in System Reliability Assessment, 1976.

2. Fussell, J. B., Henry, E. B., and Marshall, N. H., MOCUS: A Computer Program to Obtain Minimal Cut Sets From Fault Trees, ANCR-1156, 1974.

3. Lees, F. P., Loss Prevention in the Process Industries, London: Butterworths, 1980.

4. McCormick, N. J., Reliability and Risk Analysis, New York Academic Press, Inc., 1981.

5. Vesely, W. E., et al, Fault Tree Handbook, NUREG-0492, 1981.

4.9 Event Tree Analysis

Description

Event tree analysis evaluates potential accident outcomes that might result following an equipment failure or process upset known as an initiating event. Unlike fault tree analysis, event tree analysis is a "forward-thinking" process; that is, the analyst begins with an initiating event and develops the following sequences of events that describe potential accidents, accounting for both the successes and the failures of the safety functions as the accident progresses.

Event trees are a modified form of the decision trees traditionally used in business applications. Event trees provide a precise way of recording the accident sequences and defining the relationships between the initiating events and the subsequent events that combine to result in an accident. Then by ranking the accidents, or through a subsequent quantitative evaluation, the most important accidents are identified.

Event trees are well-suited for analyzing initiating events that could result in a variety of effects. An event tree emphasizes the initial cause and works from the initiating event to the final effects of the event. Each branch of the event tree represents a separate effect (event sequence) that is a clearly-defined set of functional relationships.

Guidelines for Using Procedure

The general procedure for event tree analysis contains four steps:

1. Identify an initiating event of interest
2. Identify the safety functions designed to deal with the initiating event
3. Construct the event tree
4. Describe the resulting accident event sequences.

Each of these steps is discussed below with the aid of an example.

1. Identify the Initiating Event

Selecting the initiating event is an important part of event tree analysis. This event should be a system or equipment failure, human error, or a process upset that can result in a variety of effects depending on how well the system or operators respond to the event. If the selected event results directly in a specific accident, it is better-suited for a fault tree analysis that will determine its causes. In most applications of event tree analysis, the initiating event is "anticipated"; that is, the plant design includes systems, barriers, or procedures that are intended to respond to and mitigate the effects of the initiating event.

For our example, we define our initiating event as "Loss of cooling water to an oxidation reactor."

2. Identify the Safety Functions Designed to Deal with the Initiating Event

The safety functions (safety systems, procedures, operator actions, etc.) that respond to the initiating event can be thought of as the plant's defense against the occurrence of the initiating event. These safety functions usually include:

- Safety systems that automatically respond to the initiating event (including automatic shutdown systems)
- Alarms that alert the operator when the initiating event occurs
- Operator actions designed to be performed in response to alarms or required by procedures
- Barriers or containment methods that are intended to limit the effects of the initiating events.

In particular, these safety functions influence the ultimate effects of any accident resulting from the initiating event. The analyst should identify all safety functions that can change the result of the initiating event in the chronological order that they are expected to respond. The descriptions of these safety functions should state their intended purpose. The successes and failures of the safety functions are accounted for in the event tree.

For our example, we consider the following safety functions in response to the initiating event "Loss of cooling water to oxidation reactor":

- Oxidation reactor high temperature alarm alerts operator at temperature T1
- Operator reestablishes cooling water flow to the oxidation reactor
- Automatic shutdown system stops reaction at temperature T2.

These safety functions are listed in the order in which they are intended to occur. This example assumes that the alarm and shutdown system each have their own temperature sensor and that the temperature alarm is the only alarm that alerts the operator to the problem.

3. Construct the Event Tree

The event tree displays the chronological development of accidents, beginning with the initiating event and proceeding through successes and/or failures of the safety functions that respond to the initiating event. The results are clearly-defined accidents that can result from the initiating event.

The first step in constructing the event tree is to enter the initiating event and safety functions that apply to the analysis. The initiating event is listed on the left-hand side of the page and the safety functions are listed, in chronological order, across the top of the page. Figure 4-16 shows this completed step for our example. The line underneath the initiating event description represents the progression of the accident path from the occurrence of the initiating event up to the first safety function.

The next step is to evaluate the safety function. Only two possibilities are considered: success of the safety function, and failure of the safety function. This evaluation assumes that the initiating event has occurred. The analyst must decide whether the success or failure of the safety function affects the course of the accident. If the accident is

SAFETY
FUNCTIONS:

Oxidation reactor
high temperature
alarm alerts operator
at temperature T1

Operator
reestablishes
cooling water flow
to oxidation reactor

Automatic
shutdown system
stops reaction at
temperature T2

INITIATING EVENT:

Loss of cooling water
to oxidation reactor

FIGURE 4-16.   FIRST STEP IN CONSTRUCTING EVENT TREE FOR EXAMPLE CITED IN TEXT

affected, a branch point is inserted in the event tree to distinguish between the success and the failure of the safety function. Normally, success of the function is denoted by an upward path and failure of the function by a downward path. If the safety function does not affect the course of the accident, the accident path proceeds, with no branch point, to the next safety function.

For our example, the first safety function (high-temperature alarm) does affect the course of the accident by alerting the operator, as shown by the branch-point development in Figure 4-17.

Every branch point developed in the event tree creates additional accident paths that must be evaluated individually for each of the subsequent safety systems. When evaluating a safety function on an accident path, the analyst must assume the conditions (previous successes and failures) dictated by the path of the accident up to the safety function. This can be seen in our example when evaluating the second safety function (Figure 4-18). The upper path requires a branch point because the high-temperature alarm (first safety function) was successful and the operator can affect the course of the accident by success or failure in performing. The lower path has no branch point because the high-temperature alarm failed and the operator was not alerted to the effects of the initiating event and therefore had no opportunity to affect the course of the accident. The lower accident path proceeds directly to the next safety function.

Figure 4-19 shows the completed event tree for our example problem. The uppermost accident path has no branch point for the automatic shutdown system (third safety function) because both the alarm and the operator were successful and the shutdown system is not required. The other accident paths contain branch points for the shutdown system because it can affect the outcome of the accident paths.

4. Describe the Accident Sequence

The last step of the event tree analysis procedure is to describe the accident event sequences. The sequences will represent a variety of outcomes that can follow the initiating event. One or more of the sequences

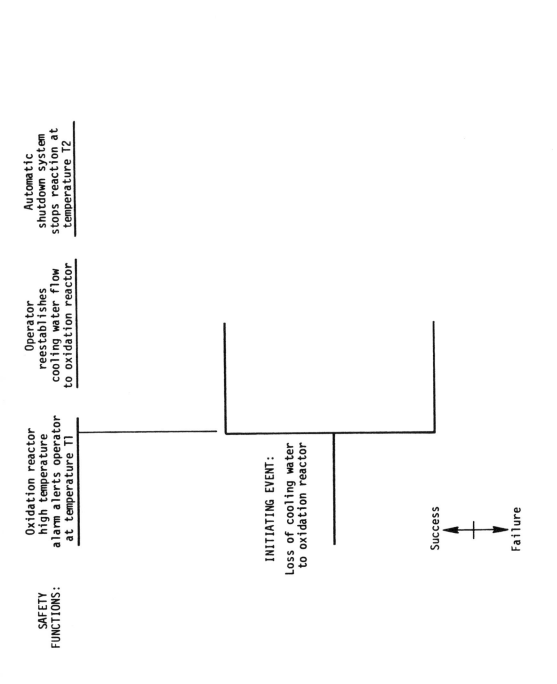

FIGURE 4-17. REPRESENTATION OF THE FIRST SAFETY FUNCTION IN THE EVENT TREE FOR THE EXAMPLE CITED IN TEXT

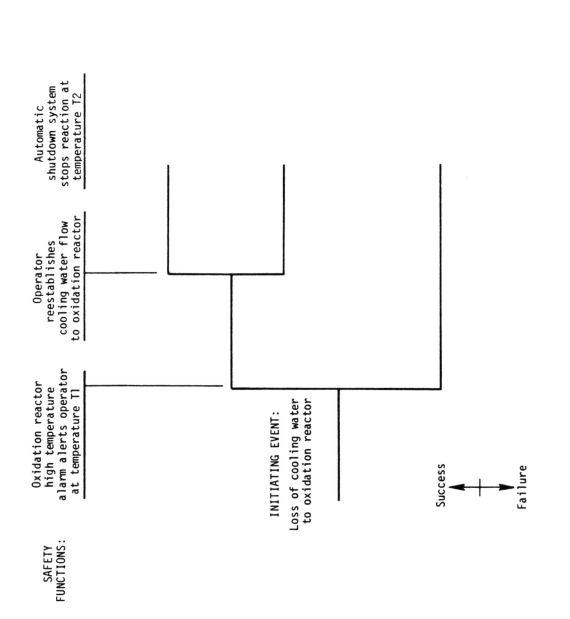

FIGURE 4-18. REPRESENTATION OF THE SECOND SAFETY FUNCTION IN THE EVENT TREE FOR THE EXAMPLE CITED IN TEXT

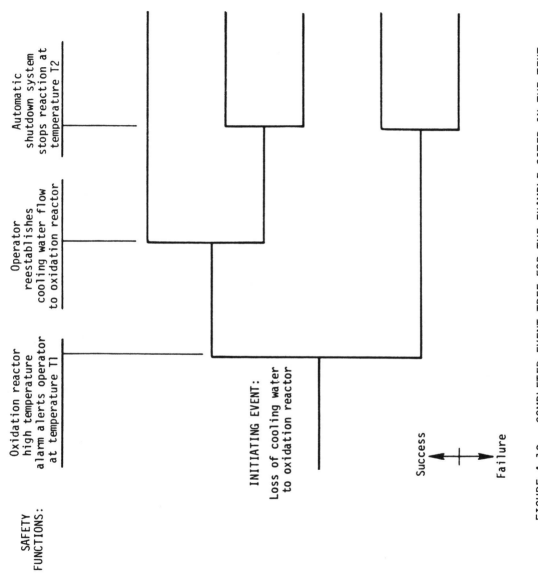

FIGURE 4-19. COMPLETED EVENT TREE FOR THE EXAMPLE CITED IN THE TEXT

may represent a safe recovery and a return to normal operations or an orderly shutdown. The sequences of importance, from a safety viewpoint, are those that result in accidents.

The analyst should examine the successes and failures in each resulting sequence and provide an accurate description of its expected outcome. This description should be as detailed as necessary to describe the accident. Figure 4-20 has sequence descriptions for the example entered for each accident path in the event tree. Figure 4-20 also shows a shorthand notation often employed in event tree analysis. Each event in the diagram is assigned a symbol (in this case A, B, C, and D) and the sequences are represented by the symbols of the events that fail and cause that particular accident. For example, in Figure 4-20 the uppermost sequence is simply labeled "A". This sequence is interpreted as "the initiating event occurs with no subsequent failures of the safety functions." While this shorthand method is useful for display purposes, it should not be substituted for a complete sequence description.

Once the sequences are described, the analyst can rank the accidents based on the severity of their outcomes. The structure of the event tree, clearly showing the progression of the accident, helps the analyst in specifying where additional procedures or safety systems will be most effective in protecting against these accidents.

If data are available, a quantitative analysis that estimates accident probabilities from the individual event probabilities can provide additional information for ranking the accidents and determining priorities for additional protective measures.

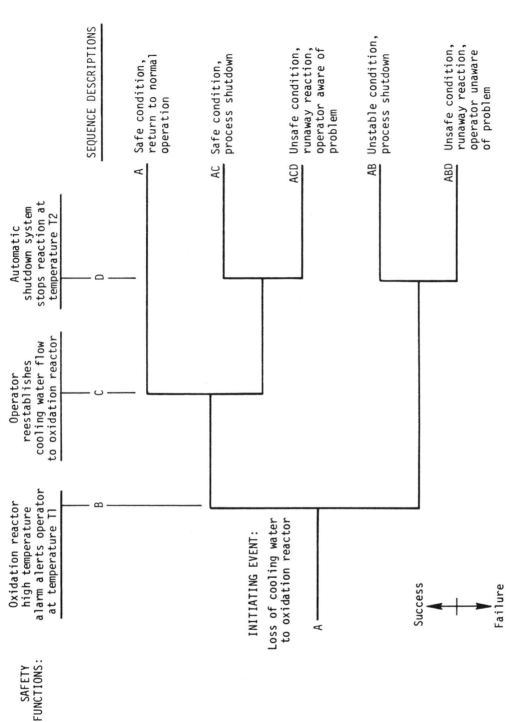

FIGURE 4-20. ACCIDENT SEQUENCES DERIVED FROM COMPLETED EVENT TREE

References

1. Atomic Energy Commission, Reactor Safety Study, WASH-1400, 1975.

2. Lees, F. P., Loss Prevention in the Process Industries, London: Butterworths, 1980.

3. McCormick, N. J., Reliability and Risk Analysis, New York, Academic Press, Inc., 1981.

4.10 Cause-Consequence Analysis

Description

Cause-consequence analysis combines the "forward-thinking" features of event tree analysis with the "reverse-thinking" features of fault tree analysis. The result is a technique that relates specific accident consequences to their many possible basic causes. Its advantage to the analyst is that it uses a graphical method that can proceed in both directions: forward toward the consequences of the event and backward toward the basic causes of the event.

The result is a cause-consequence diagram that displays the relationships between accident consequences and their basic causes. The solution of the cause-consequence diagram for a particular accident sequence is a list of accident sequence minimal cut sets. These sets are analogous to fault tree minimal cut sets because they represent all the combinations of basic causes that can result in the accident sequence. A quantitative analysis using these sets can provide estimates of the frequency of occurrence of each accident event sequence.

Guidelines for Using Procedures

A general procedure for cause-consequence analysis contains six steps:

1. Select an event to be evaluated
2. Identify the safety functions (systems, operator actions, procedures, etc.) that influence the course of the accident resulting from the event
3. Develop the accident paths resulting from the event (event tree analysis)
4. Develop the event and the safety function failure event to determine basic causes (fault tree analysis)
5. Determine the accident sequence minimal cut sets
6. Rank or evaluate results of the analysis.

1. Step 1: Select an Event to be Evaluated

The events analyzed in cause-consequence analysis can be defined in
two ways:

- TOP event (as in a fault tree analysis)
- Initiating event (as in an event tree analysis).

The event of interest is, therefore, a fault event or equipment failure that
would otherwise be suited for a fault tree or event tree analysis. Additional
information for selecting these events can be found in the appropriate
sections of the fault tree and event tree procedures (Sections 4.8 and 4.9,
respectively).

2. and 3. Steps 2 and 3: Identify Safety Functions and Develop Accident
Paths

Steps 2 and 3 parallel an event tree analysis; that is, the various
accident paths are constructed based on the chronological successes and
failures of the appropriate safety functions. The primary difference between
event tree and cause-consequence analysis is the symbols used in the diagram.
Figure 4-21 shows the symbol most often used for the event tree branch point
in the cause-consequence diagram. This branching operator contains the
function description normally written over the event tree branch point.
Figure 4-22 shows the symbol used in a cause-consequence diagram to represent
the resulting consequence. No corresponding symbol is normally used in the
event tree diagram.

4. Step 4: Develop the Event and the Safety Function Failure Event to
Determine Basic Causes

This step is actually the application of fault tree analysis to the
initiating event and safety function failure events represented in the event
tree portion of the cause-consequence diagram. Each failure of a function in
a branching operator is treated as the TOP event of a fault tree. Section 4.8
provides procedures for fault tree construction.

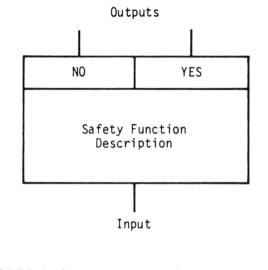

FIGURE 4-21.   BRANCH-POINT SYMBOL USED IN
                CAUSE-CONSEQUENCE ANALYSIS

FIGURE 4-22.   CONSEQUENCE SYMBOL USED IN
                CAUSE-CONSEQUENCE ANALYSIS

5. Step 5: Determine the Accident-Sequence Minimal Cut Sets

The accident sequence minimal cut sets are determined in a manner very similar to determining fault tree minimal cut sets. The accident sequence is composed of a sequence of events, each of which is a TOP event for a fault tree that is part of the cause-consequence diagram. For the accident sequence to occur, all of the events in the sequence must occur. An accident sequence fault tree is constructed by connecting all the safety function failures with an AND logic gate, with the accident sequence occurrence as the new TOP event. The standard fault tree solution technique discussed in Section 4.8 can then be used to determine the accident sequence minimal cut sets. This process can be repeated for all accident sequences identified in the cause-consequence analysis.

6. Step 6: Rank or Evaluate the Results of the Analysis

Evaluating the results of the cause-consequence analysis is a two-step process. First, the accident sequences are ranked based on their severity and importance to plant safety. Then, for each important accident sequence, the accident sequence minimal cut sets can be ranked to determine the most important basic causes. Section 4.8 discusses methods for ranking minimal cut sets.

References

1. Lees, F. P., Loss Prevention in the Process Industries, London: Butterworths, 1980.

2. Nielsen, D. S., "Use of Cause-Consequence Charts in Practical Systems Analysis", in Reliability and Fault Tree Analysis, SIAM, 1975.

3. Taylor, J. R., Cause-Consequence Diagrams, NATO Advanced Study Institute, 1978.

4.11 Human Error Analysis

There are several task-analysis-based qualitative techniques for performing Human Error Analysis. Many of these techniques have been in use for considerable periods of time. Early work in this area grew out of industrial engineering efficiency analyses. Human-performance and human-error evaluations were developed following World War II. In the context of safety studies, early forms of human error analysis were conducted for military, chemical, and transportation systems; quantification of human errors followed in the weapons and nuclear power industries.

These techniques are best applied in the field by professionals trained in human-performance technology. It is possible, however, to perform human error analyses using operations-oriented staff members such as training specialists.

Human error analysis, although simple to perform, is highly interactive with (or within) other types of hazard evaluation. Human error is taken into account in several of the previously mentioned hazard-evaluation procedures (in some cases explicitly, in others, implicitly).

To perform a human error analysis outside the context of an entire system analysis is to put the emphasis of the analysis on human-performance aspects to the exclusion of most equipment characteristics. This would be likely if the system in question were one known or expected to be prone to human-error-initiated accidents. In most cases, however, it is recommended that the human error analysis be performed in conjunction with a more complete hazard analysis. This can be accomplished by considering human-performance characteristics of system operation during the hazard evaluation itself or by performing a human error analysis separately and integrating the results into the hazard evaluation.

Techniques for human error analysis are part of a specialized sub-discipline of hazard evaluation and, as such, have not been described here. Information on these types of evaluations is available from several sources. The techniques themselves are readily adaptable to chemical process-plant analysis situations.

References

1. Bell, B. J. and Swain, A. D., A Procedure for Conducting Human
    Reliability Analysis, NUREG/CR-2254, U.S. Nuclear Regulatory Commission,
    Washington, D.C., May, 1983.

2. Swain, A. D. and Guttmann, H. E., Chapter 4, "Man-Machine Systems
    Analysis", in Handbook of Human Reliability Analysis with Emphasis on
    Nuclear Power Plant Applications, NUREG/CR-1278, U.S. Nuclear Regulatory
    Commission, Washington, D.C., August, 1983.

3. Meister, D., Chapter 2, "Human Error and Human Factors", and Chapter 3,
    "Methods of Performing Human Factors Analyses", in Human Factors: Theory
    and Practice, John Wiley & Sons, New York, New York, 1971.

APPENDIX A

A.1 SAMPLE HAZOP CHECKLIST

1. Assess Consequences of a Tube Leak

    • Pressure (where HP side DP exceeds 1.5 LP side DP)

    • Temperature (including auto-refrigeration)

    • Chemical reaction
      - overpressure
      - overtemperature
      - formation of solids

    • Is pressure relief path adequate and open?

    • Other consequences of
      - leakage of toxics/flammables to
        undesirable location
      - corrosion, embrittlement or similar effect

    • Where double-pipe HP-to-LP differential $\geq$ 1000 psi

2. Can Design Pressures be Exceeded?

    • Maximum pressure source - upstream or downstream

    • Pressure drop between reboiler and PR valve

    • Path of escape or relief adequate and open?

    • Adequacy of sizing of PR valve
      - external fire contingency included for
        reboiler of >1000 gallon liquid

    • Shutoff against maximum pressure source
      - upstream, downstream, thermal expansion
        (trapped cold side)

    • Can vacuum be created?

3. Can Design Temperatures be Exceeded?

    • Maximum upstream temperature at source

    • Bypassing of upstream heat removal equipment

    • Loss of flow on cold side

4. Loss of Level

    • Blow-through of vapor to next vessel

(kettle-type boilers only)

5. Is Temperature too High for Downstream Tank?
    (coolers only)

    ● THA in rundown line to warn of loss of cooling?

PUMPS

Centrifugal

1. Can Casing Design Pressure be Exceeded?

    ● Maximum suction pressure + shutoff delta P
      - Note: If pump curve not available, shut-off delta P (motor
        drive) may be estimated as 120% of operating delta P and shut-
        off delta P (turbine-drive) may be estimated as 132% of
        operating delta P.

    ● Higher than design specific gravity of pumped fluid

      - during startup
      - during upsets

2. Is Downstream Piping/Equipment Adequately Rated?

    ● If downstream blockage raises suction pressure:
        DP = maximum suction P + shutoff delta P

    ● If downstream blockage does not raise suction
      pressure,
        DP = the greater of:
            normal suction plus shutoff delta P
                        or
            maximum suction plus normal delta P

3. Is Backflow Prevented?

    ● Check valve in discharge

    ● Double check valve for delta P >1000 psi

4. Suction Piping Overpressure (single pumps)

    ● Suction valve, flange and connecting piping
      same DP as suction line DP

5. Suction Piping Overpressure (parallel pumps)

   - Suction valve and intervening component DP
     > 3/4 pump discharge DP

6. Is Damage from Low Flow Prevented?

   - Recycle system to ensure 20% best-efficiency-point flow

7. Can Fire be Limited?

   - Provide isolation valve(s) if suction vessel(s)
     inventory is:
     - light ends > 2000 gallons, or
     - HC liquid above 600°F >2000 gallons, or
     - HC liquid > 4000 gallons

   - Remote actuated if
     - 10" or larger line size, or
     - located in fire risk area (< 25 ft. horizontal)
       Note: Valve activator and exposed cable fireproofed

Positive Displacement

1. Can Casing Design Pressure be Exceeded?

   - PRV in discharge
   - Set pressure = casing DP minus maximum suction P
   - PRV discharge location (viscous materials)

CENTRIFUGAL COMPRESSORS

1. Can Design Pressure(s) be exceeded (any component)?

   - Backflow due to trip-out. Minimum DP must be
     suitable for settling pressure.

   - Backflow via recycle loop
     - Check valves to protect against backflow from
       downstream
     - Restriction to limit recycle flow and safety
       valve protection of low pressure stages sized
       for maximum recycle

   - Abnormally high molecular weight

   - Overspeed

2. Can Design Temperature be exceeded?

- Too little flow (surge)
  - Blocked outlet
  - High downstream pressure
  - Too low molecular weight

- Loss of cooling
  - Upstream
  - Interstage

3. Is there other chance of mechanical damage?

- Liquid carryover
  - Adequate suction KO drum w/alarms, cutouts?
  - Is traced suction line required?

- Surge
  - Adequate automatic recycle?
  - Recycle inside RBVs?

- Reverse rotation
  - Check valves each stage discharge?

- Air entry into machine
  - Can abnormal conditions permit vacuum?
    If so, is system designed to accommodate?

- Excessive Speed
  - Is overspeed cutout provided?

4. Can fire be limited?

- Remote shutdown if > 200 BHP

- Isolation valves
  - Total liquid holdup in KO drums <1000 gallon or removable in 10
    minutes: - suction, discharge & recycle RBV's required.
  - Total liquid holdup > 1000 gallon not removable in 10
    minutes: - individual RBV's in each stage required.

5. Consideration for SV(s)

- Blocked discharge @ maximum suction pressure

- Loss of cooling

- Overspeed - to 105% design or at trip on
  turbine drives

- Molecular weight above design

POSITIVE DISPLACEMENT COMPRESSORS

1. Can design pressure(s) be exceeded (any component)?

   ● Backflow through recycle loop
     - SV's for low pressure stages sized for maximum recycle
     - Consideration of parallel machines
     - Restriction to limit recycle flow - high pressure machines

   ● Shutoff condition in discharge (SV usually required).

2. Can design temperature be exceeded?

   ● Loss of cooling
     - Feed or recycle gas
     - Cylinder jacket

   ● Running on total recycle (uncooled)

   ● Consider exothermic decomposition of compressed fluid

3. Is there chance of mechanical damage?

   ● Liquid carryover
     - Adequate suction KO drum
     - Traced suction line

   ● Air entry into machine
     - Can abnormal conditions permit vacuum? If so, is system designed to accommodate?

4. Can fire be limited?

   ● Remote shutdown if > 200 BHP

   ● Isolation valves
     - Total holdup in KO drums < 1000 gallon or removable in 10 minutes: - Suction, discharge & recycle RBV's required.
     - Total liquid hold > 1000 gallon not removable in 10 minutes: - Individual RBV's in each stage required.

CONTROL VALVES (CVs)

- On failure of control medium or signal, does the CV fail in a way to:
  - reduce heat input (cut firing, reboiling, etc.)?
  - increase heat removal (increases reflux, quench, perhaps feed, etc.)?
  - maintain or increase furnace tube flow?
  - assure adequate flow at compressors or pumps?
  - reduce or stop input of reactant?
  - reduce or stop makeup to a recirculating system?
  - bottle the unit in?
  - avoid overpressuring upstream or downstream equipment?
  - avoid overcooling (below minimum desired temperature)?

- Upon plant-wide or single CV failure of control medium or signal, are conflicts among the above-listed objectives satisfactorily resolved?

- Is there provision in the design for failure at a single CV in which the valve either
  - sticks in the opposite from the design failure position, or, where applicable,
  - sticks in the single other position which is considered most undesirable?
  - fails open with bypass 50% open?

- Will any mode of CV failure result in overpressuring or other risk to equipment or piping, downstream or upstream? For example:
  - are upstream vessels between a source of pressure and the CV designed for the maximum pressure when the CV closes?
  - where piping class decreases after the CV, is the piping suitable for the CV wide open with downstream block closed? What about vessels or other equipment in the same circuit?
  - is there any equipment whose material selection makes it subject to rapid deterioration or failure on any specific maloperation or failure of the CV (overheating, overchilling, rapid corrosion, etc.)?
  - will reactor temperature run away?
  - does the threeway valve used in a pressure-relieving path have full path opening in all valve positions?

- Is on-line testing provided for safety cut-ins, cut-outs, or supplementary alarms? For any such safety system:
  - should failure or manual deactivation of the sensing-signal-control system be shown by an alarm?
  - is blowback or purge advisable on the sensing system or CV?

- should manual bypassing of the CV be prevented by
  car-seal?

TOWERS AND PROCESS-RELATED DRUMS

1. Limiting Fuel to a Fire or Emergency

   - Are emergency isolation valves provided in:
     - lines to suction and from discharge of
       200 HP$^+$ compressor?
     - drum lines to/from 200$^+$ HP compressor interstages
       (> 1000 gallon)?
     - lines to pumps where vessel maximum working inventory
       plus tray holdup is:
           over 2,000 gallon light ends, or
           over 2,000 gallon HC liquid above 600°F, or
           over 4,000 gallon HC liquid
     - 2 inch and smaller liquid lines from vessel containing 1,000-
       10,000 gallon light ends or material
       above its flash point?
     - all lines below maximum liquid level on vessel
       containing over 10,000 gallon HC liquid?
     - liquid lines in congested areas or at vessels
       with sub-standard spacing?
     - lines feeding equipment exceptionally vulnerable
       to fracture (e.g., Karbate exchangers)?

   - Are emergency isolation valve and driver type and
     installation in accordance with appropriate practices?

   - Should vapor blowdown or liquid pulldown option
     be applied?

2. Overpressure

   - Is PRV protection provided for the controlling
     contingency, including utility failure, external
     fire, and operating failure?

   - Is thermal relief required on a small liquid-filled
     vessel which needs no fire or other PRV?

   - Can failure-open of an automatic controller
     overpressure a downstream vessel?

   - Is path to PRV free of obstructions?
     - avoid CWMS in path
     - potential restriction by catalyst, coke,
       refractory, reaction products?

- Does remote but possible contingency, e.g., CSO closure, not exceed 150% DP?

- Is water separation and drawoff required to prevent delivering water to another vessel where it could vaporize at excessive rate?

3. Collapse

- Does the vessel require design for vacuum?
  - Is the atmospheric pressure boiling point of the processed material above ambient temperature? (vacuum on cooldown)

4. Reducing Leakage Risk

- Are there double valves on regularly used light ends drains/sample points?

- Are drains on vessels containing auto-refrigeration liquid double-valved, with quick opening valve nearest vessel?

- Are other drains and vents suitably valved and blinded/plugged?

- Is winterizing required because water can accumulate/freeze?

5. Furnace Fuel Gas Drum

- Does it have proper instrumentation and provision for clearing liquid? (see Furnace Checklist)

- Is the drum-to-burner line heat-traced?

(FIXED BED) SINGLE PHASE REACTORS
AND STIRRED REACTORS

1. Overpressure

- Is PRV (Pressure Relief Valve) protection provided for the controlling contingency, including utility failure, external fire, and operating failure?

- PRV inlet subject to plugging?

- Are design delta P's taken into account in setting remote PRV versus reactor DP?

- Path to remote PRV subject to fouling increasing delta P over design?

- Bed subject to plugging or blockage?
  - Scale accumulation (external source)?
  - Coking or other solid reaction byproduct?
  - Catalyst attrition?
  - Support failure (due above design delta P) and bed shift to restrict outlet?

- Is air properly isolated from reactor when not used in regeneration? (overpressure/overtemperature)

- Would leakage of coolant from internal coil into reactor overpressure it?

2. Overtemperature (Often with accompanying Overpressure)

- Excessive preheat?

- Exothermic reaction?
  - Quench failure or loss of external cooling?
  - Excess or deficiency of one reactant?
  - Can loss of agitation in cooled stirred reactor lead to excessive temperature/pressure?
  - Could loss of agitation in heated jacketed reactor lead to localized overheating at liquid surface and subsequent runaway?
  - Local hot spot due to partial bed obstruction?
  - Excessive point or surface temperature lead to thermal decomposition or runaway?
  - Delayed onset of batch reaction while continuing reactant addition?
  - Would leakage into reactor of coolant from jacket or internal coil react exothermically?
  - Could backflow of a reactant through a depressuring system lead to or exacerbate a runaway?
  - Excessive preheat drives reaction further?
  - Appropriate normal control/emergency instrumentation provided?
  - Provision for on-line test of emergency cut-off/ dump/isolation?

- Regeneration (in place)
  - Maximum regeneration temperature provided for?
  - Too much burn medium, i.e., too high concentration?

- (Sufficient TI/THA coverage in beds?)

3. <u>Degradation</u>

   ● Vessel materials attacked at excessive rate during regeneration or abnormal reaction?

4. <u>Fire</u>

   ● Is dumped catalyst pyrophoric? In-situ deactivation required?

5. <u>Toxicity</u>

   Is atm. vent gas acutely toxic at any stage of regeneration or batch reaction?

   ● Are there any acutely toxic emissions from dumped catalyst, regenerated or not?

FURNACE - FIREBOX SIDE

Protection Against Liquid Entry Via the Fuel Gas System

● Is fuel gas KO drum (uninsulated) provided for fuel gas/pilot gas/waste gas systems?

● Is manual block valve accessible at least 50 feet from furnace on each fuel?

● Is provision made for draining liquids from KO drum (preferably to closed system)?

● Does drain need backflow protection?

● Is KO drum equipped with level gauge and LHA?

● Is fuel line heat-traced/insulated from drum to burner?

Protection Against Firebox Explosion

● Does safety instrumentation protect against pilot flameout? Does PLCO on pilot gas cut pilot & all other fuels, or does flame scanner cut out all fuels?

● Are all cutout valves single-seated and dedicated?

● Do all cutout valves require manual reset?

● Is provision made for regular testing of the cutout systems (bypass circuit or short run lengths)?

- Are all cutout valve bypasses car-sealed closed?

- Are PHA and PLA provided on both fuel gas and pilot gas downstream of control valve?

- Are steam purge connections provided for startup? (Is valve 50 feet away?)

- Is provision made for damper failure? Fail-safe damper?

- PHA in firebox (forced draft furnaces only)?

Minimizing Consequences of a Fire

- Are toe walls provided (liquid feed or fuel)?

- Are steam purge connections provided? Is smothering steam provided to header boxes (if plug headers used)? (Is valve 50 feet away?)

FURNACE - PROCESS SIDE

Protection Against Tube Failure

Loss of process flow

- Is FLA and FLCO provided to stop flow of fuel (not pilots)?

- Is individual pass flow control provided (not all-vapor)?

- Are individual pass FLA's provided (not all-vapor)?

- Are explosion hatches provided on firebox for >1000 psi?

- Is check valve provided in coil outlet?
  - >200 psi (clean service)

- Is RBV provided for coil outlet (e.g., Powerformer or P >1000 psi)?

- Is safety valve provided in coil outlet of RBV?
  - steam purged if coking service

- Is provision made for manual cutoff of feed to furnace at least 50 feet away?

PRESSURE RELIEF VALVES (PRVs)

1. ### Application Concerns

   - Is the PRV protection adequate for the controlling contingency, including utility failure, external fire, and operating failure?

   - Is at least one PRV set at/below DP of protected equipment?

   - Should two PRVs with staggered settings be considered to avoid chattering (where alternate probable contingency releases under 25 percent of maximum capacity)?

   - Is PRV materials selection consistent with corrosiveness, auto-refrigeration, etc.?

   - Is heat tracing required to avoid inlet plugging by congealing/freezing?

   - Is blowback required or should an upstream rupture disc (RD) be used to keep coke or solids from accumulating in PRV inlet?

   - Is balanced bellows PRV avoided in congealing or viscous services?

   - Is balanced bellows vent routing flagged for dangerous materials?

   - Should a rupture disc (RD) be considered because of 1000 psig set pressure or impact-type pressure buildup?

2. ### PRV With Upstream RD

   - Is space between RD and PRV continuously vented to atmosphere?
     - If vent contains excess-flow check valve, is a monitor PI provided?

   - Is downrating of PRV capacity consistent with code?
     - uncertified RD downrates PRV to 80 percent capacity

3. ### Piping

   - Do inlet and outlet lines have same or larger flow area than PRV inlets and outlets?

   - Are inlet and outlet line ratings consistent with the PRV nozzle ratings?

   - Is any maintenance isolation valve on PRV inlet or outlet CSO'd?

   - Are CSO'd block valves avoided in circuits protect by PRV? If unlikely closure of such CSO, would pressure not exceed 150 percent DP?

- Is PRV inlet line delta P no more than 3 percent set pressure? Especially review under 50 psig set pressure or complex piping.

- Is maximum built-up back-pressure at PRV under 10 percent set pressure? (Balanced bellows under 50 percent set pressure?) (Note capacity correction required.)

- Is maximum superimposed back-pressure at PRV under 25 percent set pressure? (Balanced bellows under 75 percent set pressure?) (Note capacity correction required.)

4. Atmospheric Releases

- Is possible liquid release prevented by 15 minute holdup above LHA?

- Is outlet line velocity under 75 percent sonic?

- Is (1") snuffing steam connection to riser required?
  - 600 F + HC or hydrogen or methane service.

- Is toroidal ring outlet specified for hydrogen or methane service?

- Would ignition of release give over 6000 Btu/hr ft^2 to ground?
  - increase height or tie into closed system.

- Would ignition of release of hydrogen, methane, ethane, ethylene or HC above 600 F give over 3000 Btu/hr ft^2 to ground?
  - add automatic instrumentation to reduce release rate or probability.

PIPING

1. Pressure-Temperature Rating

- Is pipe class rating suitable for continuous conditions?

- If pipe class normal rating can be expected occasionally, does this meet the ANSI piping code limitations on
  - intermediate time: 120 percent normal press. rating, 50 hours at one time, 500 hours/year?
  - short time: 133 percent normal press. rating, 10 hours at one time, 100 hours/year? e.g., result of CV failure, pump shutoff, valve closure.

- Does any line warrant specification as a Special Line because there are short or intermediate time conditions/limitations that affect piping detail design or operating procedures?

- If piping is protected by a PRV, does the PRV setting allow for static head and flow delta P to the PRV?

• Are all valves, including both double valves and piping between them, designed for the more severe of the connecting line classifications?

• Is piping normal rating on pump discharge downstream of the pump block valve(s) suitable for the greater of:
  - normal pump suction pressure + 120 percent of normal pump delta P? (132 percent on turbine-driven pump) or maximum pump suction pressure + normal pump delta P?

• Are suction valves and downstream suction piping on parallel pumps suitable for 75 percent of discharge DP at DT?

• Where there is an alternate source of pressure at least equal to pump discharge pressure in the discharge line of a single pump, does the pump suction valve and downstream piping rating equal discharge line rating?

• Would the pressure not exceed 150 percent DP at DT in the unlikely event of closure of a CSO valve?

2. Safety and Closed Drain System

• Are blowdown lines untrapped and sloped at least 1:480 continuously downward toward the blowdown drum?

• Are PSV release connections into blowdown lines specified to avoid risk of acoustic failure?

• On closed drain headers for flammable liquids
  - Are check valves provided at each drained vessel that could be over-pressured by other drainage into a blocked closed drain?
  - Does the header rating satisfy the highest pressure rating of equipment tied into it? (Alternate is PRV on header.)
  - Is the header material suitable for the maximum and minimum temperature resulting from discharging into it (including auto-refrigeration)?
  - Would raising to ambient temperature of a blocked closed drain system require uprating or a PRV? Also consider effect of heat tracing, if applicable.
  - Is heat tracing required because the system may receive heavy solidifiable material or water (or moisture) which may freeze?

3. Safety-Oriented Provisions

• Are 10 inch and larger emergency isolation valves (see Tower and Drum Checklist) motor operated?

• With back leakage through a check valve would the pressure stay with 150 percent of DP?

• Where a restriction orifice (special case) or a CV is used as a means of limiting the capacity of a pressurization path, are all the conditions satisfied?

• Are utility connections made properly?
  - Are potentially hazardous connections made by breakaway, swing elbow, etc., where consistent with operating frequency? If not feasible, is suitable blinding and blind-isolating provision made?
  - Are check valves in the downstream hard pipe for breakaway connections into process?

• Are double valves required because DT is over 1000 F piping rating class is over 600? Or slurry service 400 F or above?

• Is a bleed valve (plugged) shown between double block valves in piping?

• Are light end sample outlets double valved?

• Are all tower and drum vents and drains specified?
  - Are their ratings consistent with the vessel DP/DT?
  - Are all drains valved and, where required, plugged, capped, or blinded?
  - Are double valves provided on regularly used drain connections for vessels?
  - Are drains on vessels containing auto-refrigeration liquid double-valved, with quick-opening valve nearest vessel?
  - Are vents not normally or frequently open plugged, capped or blinded and, where required, also valved?
  - Is there a 6 inch or larger vent (or vent capability) on all vessels in which manual entry is planned?

4. Other Areas

Is there any special consideration, either normal or connected with short-time conditions, that could promote or pre-dispose the piping to failure? For example:

• Auto-refrigeration of light ends
  - Is materials selection and piping mechanical design specified for lowest temperature during startup and shutdown as well as normal operation?
  - Does any other credible operation or contingency set the controlling (minimum) DT?

• Freezing of accumulated water
  - Is heat tracing specified for piping where water freezing is possible due to accumulation or intermittent service in cold weather?

• Heat tracing selection
  - Is the tracing temperature, possibly together with the pipe material, able to promote:

+ exothermic decomposition? (Ethylene?)
+ blockage of a pressure relief path by coking or by dryout of
  settled slurry?
+ rapidly corrosive chemical reaction?
- If so, is the method of tracing or the medium specified in a way to
  preclude? How positive?

● Stress corrosion
  - Is C.S. specified to be stress-relieved for maximum temperature vs.
    caustic concentration?
  - If heat tracing of caustic piping is required, is the maximum trac-
    ing temperature defined? How positive is the method of tracing
    temperature control?
  - Damp salt air protected against on austentic steel?

● Acid attack
  - Materials selection vs. sulfuric acid concentration/temperature
    control? (special concern when acid is mixed with HC or enters
    vessel containing HC's).
  - Material selection vs. other mineral acid or acidic-organic
    compounds?

● Metal dusting
  - For pipe running at over 900 F with high H_2, CH_4, or CO content, is
    suitable provision made (e.g., sulfide addition) to prevent catas-
    trophic failure by metal dusting?

● Erosion
  - Has sound provision been specified for erosive services? Special
    concern where high velocity HC vapor stream carries or may contain
    abrasive solids.

● Valve closure damping
  - Should a maximum closing rate be specified for motor-operated valves
    or rapid-closing hand valves whose closing could cause liquid
    hammer?
  - Is a closure damping device required on major check valves to avoid
    damage to rotating equipment? If so, does the damping system
    require specification and possible testing facilities to assure the
    check valve's performing its basic function under conditions
    unfavorable to the damping system?

A.2 SAMPLE SAFETY REVIEW CHECK LISTS

PERSONNEL SAFETY REVIEW CHECK LIST

A. Project Site Location

1. Any exposure to or from the neighborhood from fire, explosion, noise, air and stream pollution?

2. Adequate access for emergency vehicles?

3. Any potential blockages of access roads by railroads, highway congestion, etc.?

4. Access roads well engineered to avoid sharp curves? Traffic signs provided?

B. Building and Structures

1. What standards are being followed in the design of stairways, platforms, ramps and fixed ladders?

2. Are sufficient general exit and escape routes available? Alternate means of escape from roofs provided? Is protection provided to persons along the line of the escape routes?

3. Adequate lighting provided?

4. Doors and windows hung to avoid projecting into or blocking walkways and exits?

5. Structural steel grounded?

C. Operating Areas

1. Are equipment, steam, water, air and electric outlets arranged to keep aisles and operating floor areas clear of hoses and cables?

2. Ventilation furnished for hazardous fumes, vapors, dust and excessive heat?

3. Temporary storage provided for raw materials at process points, and for finished products?

4. Where operations are potentially hazardous from the standpoints of fire and explosion, are controls housed in separate structures? If not, are control room windows kept to a minimum and glazed with laminated safety glass?

5. Are alternate escape routes to safe locations provided?

6. If needed, what type of pressure relief venting of area is furnished?

7. Do platforms provide safe clearance for safe maintenance of equipment?

8. Are nozzles and manholes sized and located for safe cleanout, maintenance operations, and emergency removal of people from vessels?

9. What protection is provided to protect against contact with hot surfaces?

10. Is head clearance adequate in walkway and working areas?

11. Is power-driven equipment adequately guarded?

12. Are manually operated valves, switches and other controls readily accessible to the operator from a safe location?

13. Are vents located so that discharges, including liquids, do not endanger personnel, public or property? Are all vents above the highest liquid level possible in the system?

14. Are free-swinging hoists avoided? Are hoists equipped with safety hooks, limit switches, if motorized?

15. Are elevators equipped with shaftway door interlocks and car gate contacts? Are there safety astragals on bi-parting doors?

16. Is every effort being made to handle materials mechanically rather than manually?

17. Are emergency showers and hose-type eye baths provided?

18. Has a safe storage and dispensing location for flammable liquid drums been provided?

19. Are there at least two exits from hazardous work?

20. Where excessively noisy operations are concerned, what measures are contemplated to reduce the noise level to a safe range?

21. Is there safe exit from manufacturing offices or laboratories?

22. Are positive electrical power disconnects being installed for purposes of lockout?

D. Yard

1. Are roadways laid out with consideration for the safe movement of pedestrians, vehicles and emergency equipment?

2. Are railroad car puller control stations fully protected against broken cable whiplash? How will operator be protected from being caught between cable or rope and capstan or cable drum?

3. Are flammable liquid tank car and tank truck loading and unloading docks bonded or grounded?

4. Are safe means provided on loading platforms for access to work areas of tank cars and trucks?

5. Is protection against falling furnished for employees who work on tops of railroad cars and trucks?

6. Is safe access provided to tops of storage tanks on which persons go for contents measurement and vent maintenance?

7. Is there sufficient head clearance and good visibility in walking and working areas?

8. Is yard lighting adequate?

PROCESS SAFETY REVIEW CHECK LIST

Note: Consider the check list in terms not only of steady-state operation, but also startup, shutdown, and upsets of all conceivable types.

A. Materials

1. What process materials are unstable or spontaneously ignitible?

   a. What evaluation has been made of impact sensitivity?

   b. Has an evaluation of possible uncontrolled reaction or decomposition been made?

2. What data are available on amount and rate of heat evolution during decomposition of any material in the process?

3. What precautions are necessary for flammable materials?

4. What flammable dust hazards exist?

5. What materials are highly toxic?

6. What has been done to assure that materials of construction are compatible with the chemical process materials that are involved?

7. What maintenance control is necessary to assure replacement of proper materials, e.g., to avoid excessive corrosion, to avoid producing hazardous compounds with reactants?

8. What changes have occurred in composition of raw materials and what resulting changes are in process?

9. What is done to assure sufficient control of raw material identification and quality?

10. What hazards can be created by failure of supply of one or more raw materials?

11. What assurance is there of adequate raw material supply?

12. What hazards can occur as a result of loss of gas for purging, blanketing, or inerting? How certain is gas supply?

13. What precautions need to be considered relative to stability of all materials in storage?

14. What fire extinguishing agents are compatible with process materials?

15. What fire emergency equipment and procedures are being provided?

B. Reactions

1. How are potentially hazardous reactions isolated?

2. What process variables could, or do, approach limiting conditions for hazard?

3. What unwanted hazardous reactions can be developed through unlikely flow or process conditions or through contamination?

4. What combustible mixtures can occur within equipment?

5. What precautions are taken for processes operating near or inside the flammable limits?

6. What are process margins of safety for all reactants and intermediates?

7. What reaction rate data are available on the normal, or abnormally possible, reactions?

8. How much heat must be removed for normal, or abnormally possible, exothermic reactions?

9. How thoroughly is chemistry of the process known? (See NFPA "Manual of Hazardous Chemical Reactions".)

10. What foreign materials can contaminate the process and create hazards?

11. What provision is made for rapid disposal of reactants if required by plant emergency?

12. What provisions are made for handling impending runaways and for short-stopping an existing runaway?

13. How fully is the chemistry of all desired and undesired reactions known?

14. What hazardous reactions could develop as a result of mechanical equipment (pump, agitator, etc.) failure?

15. What hazardous process conditions can result from gradual or sudden blockage in equipment?

16. What raw materials or process materials can be adversely affected by extreme weather conditions?

17. What process changes have been made since the previous process safety review?

C. Equipment

1. In view of process changes since the last process safety review, how was adequate size of equipment assured?

2. Are any venting systems manifolded, and if so, what hazards can result?

3. What procedure is there for assuring adequate liquid level in liquid seals?

4. What is the potential for external fire which may create hazardous internal process conditions?

5. Is explosion suppression equipment needed to stop an explosion once started?

6. Where are flame arresters and detonation arresters needed?

7. In confined areas, how is open fired equipment protected from spills?

8. What safety control is maintained over storage areas?

9. In the case of equipment made of glass or other fragile material, can a more durable material be used? If not, is the fragile material adequately protected to minimize breakage? What is the hazard resulting from breakage?

10. Are sight glasses on reactors provided only where positively needed? On pressure or toxic reactors, are special sight glasses provided which have a capability to withstand high pressure?

11. What emergency valves and switches cannot be reached readily?

12. When was pertinent equipment, especially process vessels, last checked for pressure rating?

13. What hazards are introduced by failure of agitators?

14. What plugging of lines can occur and what are the hazards?

15. What provisions are needed for complete drainage of equipment for safety in maintenance?

16. How was adequacy of ventilation determined?

17. What provisions have been made for dissipation of static electricity to avoid sparking?

18. What requirements are there for concrete bulkheads or barricades to isolate highly sensitive equipment and protect adjacent areas from disruption of operations?

D. Instrumentation Control

1. What hazards will develop if all types of motive power used in instrumentation should fail nearly simultaneously?

2. If all instruments fail simultaneously, is the collective operation still fail-safe?

3. What provision is made for process safety when an instrument, instrumental in process safety as well as in process control, is taken out of service for maintenance? When such an instrument goes through a dead time period for standardization or when, for some other reason, the instrument reading is not available?

4. What has been done to minimize response time lag in instruments directly or indirectly significant to process safety? Is every significant instrument or control device backed up by an independent instrument or control operating in an entirely different manner? In critical processes, are these first two methods of control backed up by a third ultimate safety shutdown?

5. Has the process safety function of instrumentation been considered integrally with the process control function throughout plant design?

6. What are the effects of extremes of atmospheric humidity and temperature on instrumentation?

7. What gauges, meters, or recorders cannot be read easily? What modifications are being made to cope with or solve this problem?

8. Is the system completely free of sight glasses or direct reading liquid level gauges or other devices which, if broken, could allow escape of the materials in the system?

9. What is being done to verify that instrument packages are properly installed? Grounded? Proper design for the environment?

10. What procedures have been established for testing and proving instrument functions?

11. What periodic testing to check performance and potential malfunction is scheduled?

E. Operations

1. When was the written operating procedure last reviewed and revised?

2. How are new operating personnel trained on initial operation and experienced operating personnel kept up-to-date on plant operating procedures, especially for startup, shutdown, upsets, and emergencies?

3. What plant revisions have been made since the last process safety review?

4. What special cleanup requirements are there before startup and how are these checked?

5. What emergency valves and switches cannot be reached readily? What procedures are there to cope with these situations?

6. What safety precautions are needed in loading liquids into, or withdrawing them from, tanks? Has the possibility of static electricity creation been adequately taken care of?

7. What process hazards are introduced by routine maintenance procedures?

8. What evaluation has been made of the hazards of sewered materials during normal and abnormal operation?

9. How dependable are supplies of inerting gas and how easily can supplies to individual units be interrupted?

10. What safety margins have been narrowed by revisions of design or construction in efforts to debottleneck operations, reduce cost, increase capacity, or improve quality?

11. What provisions does the operating manual have for coverage of startup, shutdown, upsets and emergencies?

12. What economic evaluation has dictated whether a batch process or a continuous one is used?

F. Malfunctions

1. What hazards are created by the loss of each feed, and by simultaneous loss of two or more feeds?

2. What hazards result from loss of each utility, and from simultaneous loss of two or more utilities?

3. What is the severest credible incident, i.e., the worst conceivable combination of reasonable malfunctions, which can occur?

4. What is the potential for spills and what hazards would result from them?

G. Location and Plot Plan

1. Has equipment been adequately spaced and located to permit anti-cipated maintenance during operation without danger to the process?

2. In the event of the foreseeable types of spills, what dangers will there be to the community?

3. What hazards are there from materials dumped into sewers of neigh-boring areas?

4. What public liability risks from spray, fumes, mists, noise, etc. exist, and how have they been controlled or minimized?

ELECTRICAL SAFETY REVIEW CHECK LIST

A. Design

1. How completely does the electrical system parallel the process?

   a. What faults in one part of the plant will affect operation of other independent parts of the plant?

   b. How are instruments for a plant protected from faults or other voltage disturbances?

2. Are interlocks and shutdown devices made fail-safe?

   a. What is the need for each interlock and shutdown used?

   b. Are interactions and complications minimized?

   c. Is continued use of protective devices ensured?

   d. What requirements or standards were used in selecting the hardware?

3. How has the area NEC classification been established and hardware and techniques selected?

   a. What process details affect the classification, group, and division?

   b. What "UL approved" hardware is unavailable for this job? Does this require testing?

   c. Are any new techniques being applied on this job?

4. Is the electrical system simple in schematic and physical layout so that it can be operated in a straightforward manner? (This minimizes human error in switching for isolation and load transfer.)

5. What electrical equipment can be taken out of service for preventive maintenance without interrupting production? How?

6. How is the electrical system instrumented so that equipment operation can be monitored? Will this eliminate downtime due to equipment failures caused by unknown overloading?

7. What are the overload and short circuit protective devices?

   a. Are they located in circuits for optimum isolation of faults?

   b. What is the interrupting capacity?

    c. How are they coordinated?

    d. What instructions are furnished for field testing on installation and for testing during the life of equipment?

8. What bonding and grounding is provided?

    a. Does it protect against static buildup?

    b. Does it provide lightning protection?

    c. Does it provide for personnel protection from power system faults?

9. Is lighting adequate?

    a. Adequate for safe normal operation?

    b. Adequate for normal running maintenance?

    c. Adequate for escape lighting during power failure?

10. Is tankage grounding coordinated with cathodic protection?

11. Are power disconnects, starters, etc. accessible during mishaps?

12. Is communication provided to operate a complex safely (telephones, radios, signals, alarms, etc.)?

13. Are spacings and clearances furnished for normal traffic maintenance, and for fire fighting?

14. Is there a schedule for checking operability of interlocks?

15. Where sequence controllers are used, is there an automatic check, together with alarms, at key steps after the controller has called for a change, and is there a check together with alarms at key steps before the next sequence changes?

BOILER AND MACHINERY CHECK LIST

A. Boilers

    1. Safety Valves

        a. Are long and large vent lines supported?

        b. What drain connections are provided?

    c. Is first drum valve set to relieve boiler working pressure?

    d. Is the last drum valve set to pop at or below 103 percent of boiler working pressure?

2. Blow-off Piping

Is steel piping of next higher gauge than required used for boiler pressure, avoided sharp radius ells, and sloped all lines and drained all low points in the lines?

3. Feedwater Piping

Is the bypass around the feedwater regulator accessible from the operating level and located where the drum level gauge glass can be seen? Are electrically-driven feedwater pumps duplicated by steam-driven pumps?

4. Steam Outlet Piping

    a. Are there separate non-return and header stop valves where two or more boilers discharge into the same piping system?

    b. Is there a visible free blow and drain in piping between non-return and header stop valves?

    c. Are there condensate drain provisions for all sections of piping?

    d. Is there adequate piping expansion flexibility? How is piping supported?

5. Drum Water Level - Attended Operation

    a. Are there both high and low water alarms?

    b. Is there a low water cut-off of gas or oil burners? (If drop or loss of plant steam pressure does not jeopardize process safety.)

    c. Is gauge glass visible from feedwater regulator bypass valve?

    d. Is remote drum level gauge independent of drum level controls?

6. Drum Water Level - Unattended Operation

    a. Are high and low boiler water levels monitored?

    b. Are two independent low water level switches interlocked with gas or oil burner safety shut-off valves?

7. Gas Burner Control and Piping - General

   a. What type of plug cocks have been provided for manual shut-off service?

   b. Is there in-line strainer in gas line ahead of all regulating and safety shut-off valves?

   c. Do you provide for stable gas pressure regulation at all loads? This may require a small regulator in parallel with the full-sized regulator for start-up or low fire service.

   d. Is there a double safety shut-off and vent valve arrangement? What type of reset is there for each valve?

   e. What type of automatic fuel-air ratio control is used?

   f. Is there separate pressure regulation of pilot gas?

   g. Is safety control circuit DC, or 120v AC with the safety controls in the ungrounded circuit?

   h. Do you ensure positive, tamper-proof time period to provide minimum of 6 air changes in combustion chamber before light-off? Air flow rate during purge should be at least 70 percent of maximum capacity.

   i. Are controls or interlocks installed to prevent burner firing rate from being reduced below minimum stable flame?

   j. Are controls or interlocks installed to prevent burner light-off when insufficient combustion air flow is present?

   k. What interlock is there to assure low-fire burner light-off?

8. Additional gas burner controls and interlocks for unattended operation:

   a. Is main burner flame monitored?

   b. Are following interlocks for safety shut-down furnished?

      1. High gas pressure?

      2. Low gas pressure?

      3. Low combustion air flame?

      4. Low boiler water (double switched)?

   c. Is there flame scanner response line of 2-4 seconds?

d. Is there tamper-proof programmed light-off sequence to purge, light and prove pilot, light and prove main flame, post purge?

e. How have you set up positioning fuel-air ratio controls?

f. Is there a self-checking feature for flame scanner and flame scanner relay circuitry?

9. Are provisions made in the oil burner controls and piping for each of the following items?

a. Oil line strainer

b. Oil pressure control

c. Heater for heavy oil

d. Single safety shut-off valve

e. Start-up recirculating line for heavy oil

f. Positive fuel-air ratio control

g. Low oil pressure alarm or interlock

h. Low oil temperature alarm or interlock for heavy oil

i. Low atomizing steam pressure alarm or interlock

j. Positive purge cycle and low fire start controls

k. Interrupted pilot

10. Additional oil burner controls and interlocks for unattended operation:

a. Are interrupted and proved pilot and monitoring of main oil burner flame with interlock to close safety shut-off valve during flame failure provided?

b. Are the following interlocks in use for safety shut-down of burners?

1. Low oil temperature--for heavy oils?

2. Low oil pressure?

3. Low combustion air flow?

4. Low atomizing steam pressure?

5. Low boiler water (double switched)?

    c. Is a tamper-proof programmed light-off sequence provided?

    d. Are positioning fuel-air ratio controls used?

11. Are the following interlocks in use for pulverized fuel furnaces for shut-down of furnace?

    a. Igniter flame failure?

    b. Coal "flame-out" for changeover period from coal to oil?

    c. Loss of forced or induced draft or low combustion airflow?

    d. Excess furnace pressure?

Is there an igniter shut-off interlock on loss of igniter fuel pressure?

B. Piping and Valves

1. Were piping systems analyzed for stresses and movement due to thermal expansion?

2. Are piping systems adequately supported and guided?

3. Are piping systems provided for anti-freezing protection, particularly cold water lines, instrument connections and lines in dead-end service such as piping at standby pumps?

4. Are provisions made for flushing out all piping during start-up?

5. Are cast iron valves avoided in strain piping?

6. Are non-rising stem valves being avoided?

7. Are double block and bleed valves used on emergency inter-connections where possible cross-contamination is undesirable?

8. Are controllers and control valves readily accessible for maintenance?

9. Are bypass valves readily reached for operation? Are they so arranged that opening of valves will not result in an unsafe condition?

10. Are any mechanical spray steam desuperheaters used?

11. Are all control valves reviewed for safe action in event of power or instrument air failure?

12. Are means provided for testing and maintaining primary elements of alarm and interlock instrumentation without shutting down processes?

13. What provisions for draining and trapping steam piping are provided?

C. Pressure and Vacuum Relief

1. What provisions is there for flame arresters on discharge of relief valves or rupture discs on pressurized vessels?

2. What provisions are there for removal, inspection, and replacement of relief valves and rupture discs, and what scheduling procedure?

3. What need is there for emergency relief devices: breather vents, relief valves, rupture discs, and liquid seals? What are the bases for sizing these?

4. Where rupture discs are used to prevent explosion damage, how are they sized relative to vessel capacity and design?

5. Where rupture discs have delivery lines to or from the discs, what has been done to assure adequate line size relative to desired relieving dynamics? To prevent whipping of discharge end of line?

6. Are discharges from vents, relief valves, rupture discs, and flares located to avoid hazard to equipment and personnel?

7. What equipment, operating under pressure, or capable of having internal pressures developed by process malfunction, is not protected by relief devices and why not?

8. Is discharge piping of relief valves independently supported? Make piping as short as possible and with minimum changes in direction.

9. Are drain connections provided in discharge piping of relief valves where condensate could collect?

10. Are relief valves provided on discharge side of positive displacement pumps; between positive displacement compressor and block valves; between back-pressure turbine exhaust flanges and block valves?

11. Where rupture discs are in series with relief valves to prevent corrosion on valve or leakage of toxic material, install rupture disc next to the vessel and monitor section of pipe between disc and relief valve with pressure gauge and pressure bleed-off line. Have any rupture discs been installed on discharge side of relief valve?

12. What provisions for keeping piping to relief valves and vacuum breakers at proper temperature to prevent accumulation of solids from interfering with action of safety device are provided?

D. Machinery

1. Are adequate piping supports and flexibility provided to keep forces on machinery due to thermal expansion of piping within acceptable limits?

2. What is separation of critical and operating speeds?

3. Are check valves adequate and fast acting to prevent reverse flow and reverse rotation of pumps, compressors and drivers?

4. Are adequate service factors on speed changing gears in shock service provided?

5. Are there full-flow filters in lube-oil systems serving aluminum bearings?

6. Are there provisions for draining and trapping steam turbine inlet and exhaust lines?

7. Are there separate visible-flow drain lines from all steam turbine drain points?

8. Are driven machines capable of withstanding tripping speed of turbine drain points?

9. Are non-lubricated construction or non-flammable synthetic lubricants used for air compressors with discharge pressures of greater than 75 psig to guard against explosion?

10. What provisions are made for emergency lubrication of critical machinery during operation and during emergency shutdowns?

11. Are provisions made for spare machines or critical spare parts for critical machines?

12. Are there provisions for operation or safe shutdown during power failures?

13. Are vibration switches on alarm or on interlock for cooling tower fans provided?

FIRE PROTECTION CHECK LIST

1. If the building has enclosed walls and the construction or occupancy has combustibles, what kind of automatic sprinklers (wet or dry pipe systems) are provided?

2. If the building has open walls and the construction or occupancy has combustibles, how much water spray protection (HAD's or pilot head heat actuating systems) has been provided?

3. What existing hydrants serve the area or project? What additional ones are to be provided?

4. What fixed or portable monitor nozzles (on hydrants or separate) are provided for coverage of manufacturing facilities or storage facilities in open areas (not within open or closed wall buildings)?

5. Have the underground fire mains been extended or looped to supply additional sprinkler systems, hydrants and monitor nozzles? Dead ends should be avoided. What sectional control valves have been provided?

6. Are small hose standpipes provided inside of buildings?

7. What type, size, location and number of fire extinguishers are needed?

8. What flammable liquid storage tank protection has been provided? Foam? Dikes with drain valves outside the dike?

9. Where have total flooding or local-application carbon dioxide systems been provided?

10. Is load-bearing structural steel which is exposed to potential flammable liquid or gas fires fireproofed to a sufficient height above ground level to protect the steel? (This height varies from 30' to 35', depending on additional fire protection features.)

11. How has adequate drainage been provided to carry spilled flammable liquids and water used for fire-fighting away from buildings, storage tanks, and process equipment?

12. What protection has been provided for dust hazards?

13. What is the capacity of fire water supplies? What is the maximum fire water demand?

14. How long will supplies meet this maximum demand?

15. What is the spacing of flammable liquid storage tanks?

16. What is the estimated probable maximum loss (PML)?

17. What is the approximate "hold-up" of flammable liquids in the manufacturing equipment broken down by flash points?

18. What attention has been given to protection of process equipment from external fire?

19. Are liquid inventory tanks near or under the ground instead of elevated?

20. Is the area pad or flooring designed to conduct spill liquid away from process equipment? What facilities are provided for drainage?

21. How have major storage tanks or vessels been located to minimize hazard to process equipment in the event of rupture or burning?

22. Are all structures constructed of non-combustible materials and fire walls, partitions or barricades provided to separate important property damage values, high hazard operations and units important to continuity of production?

APPENDIX B

BIBLIOGRAPHY

General

Henley, E. J. and Kumamoto, H., Reliability Engineering and Risk Assessment, Prentice-Hall, Inc., 1981.

Lees, F. P., Loss Prevention in the Process Industries, London: Butterworths, 1980.

European Federation of Chemical Engineering, Risk Analysis in the Process Industries, Rugby, Warks., England: The Institution of Chemical Engineers, 1985.

Methodologies for Hazard Analysis and Risk Assessment in the Petroleum Refining and Storage Industry, CONCAWE Report No. 10/82.

Checklist/Safety Review

Balemans, A.W.M., et al., Check-list: Guidelines for Safe Design of Process Plants, 1st International Loss Prevention Symposium, 1974.

Hettig, S.G., "A Project Checklist of Safety Hazards", Chemical Engineering, 73(26), 1966.

King, R. and Magid, J., Industrial Hazard and Safety Handbook, London: Newnes-Butterworths, 1979.

Marinak, M. J., "Pilot Plant Prestart Safety Checklist", Chemical Engineering Progress, 63(11), 1967.

Parker, J. U., "Anatomy of a Plant Safety Inspection", Hydrocarbon Processing, 46(1), 1967.

Whitehorn, V. J. and Brown, H. W., "How to Handle a Safety Inspection", Hydrocarbon Processing, 46(4), 46(5), 1967.

Williams, D., "Safety Audits", in Major Loss Prevention, 1971.

Dow and Mond Hazard Indices

AIChE, Dow Process Safety Guide, 1974.

Dow Chemical Company, "Process Safety Manual", Chemical Engineering Progress, 62(6), 1966.

Dow Chemical Company, Fire and Explosion Index, Hazard Classification Guide, 5th Edition, 1981.

Lewis, D. J., The Mond Fire and Explosion Index Applied to Plant Layout and Spacing, 13th Loss Prevention Symposium, 1979.

"What If" Analysis

Cizek, J. G., Diamond Shamrock Loss Prevention Review Program, Canadian Society for Chemical Engineering Conference, 1982.

Hazard and Operability (HazOp) Studies

Chemical Industries Association, A Guide to Hazard and Operability Studies, London: Alembic House, 1977.

Cowie, C.T.Y., "Hazard and Operability Studies--A New Safety Technique for Chemical Plants", Prevention of Occupational Risks, Vol. 3, 1976.

Elliot, D. M. and Owen, J. M., "Critical Examination in Process Design", Chemical Engineer, London, 223, CE377, 1968.

Gibson, S. B., Reliability Engineering Applied to the Safety of New Projects, Teesside Polytechnic Course on Process Safety--Theory and Practice, 1974.

Gibson, S. B., I. Chem. E. Symposium Series, 47, 1976.

Gibson, S. B., "The Design of New Chemical Plants Using Hazard Analysis", Process Industry Hazards, 1976.

Gibson, S. B., Safety in the Design of New Chemical Plants (Hazard and Operability Studies), Loughborough University of Technology, 1976.

Gibson, S. B., Chemical Engineering Progress, 76(11), 1980.

Helmers, E. N. and Schaller, L. C., Plant/Operations Progress, Vol. 1, No. 3, 1982.

Henderson, J. M. and Kletz, T. A., "Must Plant Modifications Lead to Accidents?", Process Industry Hazards, 1976.

Himmelblau, D. M., Fault Detection and Diagnosis in Chemical and Petrochemical Processes, Amsterdam: Elsevier, 1978.

Imperial Chemical Industries Limited, Hazard and Operability Studies, Process Safety Report 2, London, 1974.

Kletz, T. A., "Specifying and Designing Protective Systems", Loss Prevention, Vol. 6, 1972.

Knowlton, R. E., An Introduction to Guide Word Hazard and Operability Studies, CSChE Conference, 1982.

Knowlton, R. E., An Introduction to Creative Checklist Hazard and Operability Studies, CSChE Conference, 1982.

Lawley, H. G., Chemical Engineering Progress, 70(4), 1974.

Lawley, H. G., "Operability Studies and Hazard Analysis", Loss Prevention, Vol. 8, 1974.

Lock, B., Hazard and Operability Studies, Loughborough University of Technology, 1976.

Rushford, R., "Hazard and Operability Studies in the Chemical Industries", Transactions N.E. Coast Institute Engineers and Shipbuilders, 93(5), 1977.

Solomon, C. H., "The Exxon Chemicals Method of Identifying Potential Process Hazards", I Chem E Loss Prevention Bulletin No. 52, August, 1983.

Failure Modes and Effects Analysis

Department of Navy, Procedures for Performing a Failure Mode and Effect Analysis, MIL-STD-1629A, 1977.

Jordan, W. E., "Failure Modes, Effects, and Criticality Analysis", Proceedings Reliability and Maintainability Symposium, American Society of Mechanical Engineers, 1972.

King, C. F. and Rudd, D. F., "Design and Maintenance of Economically Failure-Tolerant Processes", AIChE Journal, Vol. 18, 1972.

Lambert, H. E., Failure Modes and Effects Analysis, NATO Advanced Study Institute, 1978.

Ostrander, V. P., "Spacecraft Reliability Techniques for Industrial Plants", Chemical Engineering Progress, 67(1), 1971.

Taylor, J. R., A Formalisation of Failure Mode Analysis of Control Systems, RISO-M-1654, 1973.

Taylor, J. R., A Semiautomatic Method for Qualitative Failure Mode Analysis, RISO-M-1707, 1974.

Vesely, W. E., et al., Fault Tree Handbook, NUREG-0492, 1981.

Fault Tree Analysis

Andow, P. K. and Lees, F. P., "Process Computer Alarm Analysis: Outline of a Method Based on List Processing", Trans. Inst. Chem. Engrs., Vol. 54, 1975.

Andow, P. K., A Method for Process Computer Alarm Analysis, Ph.D. Thesis, Loughborough University of Technology, 1973.

Astolfi, M. Contini, S., and van den Muyzenberg, C. L., A Technique Suitable for the Analysis of Large Fault Trees on Minicomputers, Int. Conf. on Reliability and Maintainability, Paris, 1978.

Atomic Energy Commission, Reactor Safety Study, WASH-1400, 1975.

Barlow, R. E., Fussell, J. B., and Singpurwalla, N. D. (eds), Reliability and Fault Tree Analysis, SIAM, 1975.

Bennetts, R. G., "A Realistic Approach to Detection Test Set Generation for Combinational Logic Circuits", Comput. J., Vol. 15, 1972.

Boeing Company, Systems Safety Symposium, Seattle, 1965.

Browning, R. L., "Analyze Losses by Diagrams", Hydrocarbon Processing, 54(9), 1975.

Browning, R. L., "Human Factors in the Fault Tree", Chemical Engineering Progress, 72(6), 1976.

Brown, D. B., System Analysis and Design for Safety, New York: McGraw-Hill, 1976.

Burdick, G. R., et al., "Phased Mission Analysis: A Review of New Developments and an Application", IEEE Trans. on Reliability, R-26, 1977.

Carnino, A., "Reliability Techniques in Assessment of Nuclear Reactor Safety", Generic Techniques in System Reliability Assessment, Leyden: Noordhoff, 1976.

Crosetti, P. A. and Bruce, R. A., "Commercial Application of Fault Tree Analysis", Proceedings Reliability and Maintainability Conference, 1970.

Crosetti, P. A., "Fault Tree Analysis with Probability Evaluation", IEEE Transactions on Nuclear Science, NS-18, 1971.

Doering, E. J. and Gaddy, J. L., Optimal Process Reliability with Flowsheet Simulation, 11th Loss Prevention Symposium, New York: AIChE, 1977.

Esary, J. D. and Ziehms, H., "Reliability Analysis of Phased Missions", Reliability and Fault Tree Analysis, SIAM, 1975.

Fussell, J. B., Powers, G. J., and Bennetts, R. G., "Fault Trees: A State of the Art Discussion", IEEE Transactions on Reliability, R-23, 1974.

Fussell, J. B. and Vesely, W. E., "A New Methodology for Obtaining Cut Sets for Fault Trees", Transactions of the American Nuclear Society, 15, 1972.

Fussell, J. B., "Fault Tree Analysis: Concepts and Techniques", Generic Techniques in System Reliability Assessment, 1976.

Fussell, J. B., Synthetic Tree Model: A Formal Methodology for Fault Tree Construction, ANCR-1098, 1973.

Fussell, J. B., Henry, E. B., and Marshall, N. H., MOCUS: A Computer Program to Obtain Minimal Cut Sets from Fault Trees, ANCR-1156, 1974.

Garribba, S., et al., "Efficient Construction of Minimal Cut Sets from Fault Trees", IEEE Transactions on Reliability, R-26, 1977.

Haasl, D. F., Advanced Concepts in Fault Tree Analysis, System Safety Symposium, Boeing Company, 1965.

Himmelblau, D. M., Fault Detection and Diagnosis in Chemical and Petrochemical Processes, Amsterdam: Elsevier, 1978.

Houston, D.E.L., "New Approaches to the Safety Problem", Major Loss Prevention, 1971.

Kolodner, H. J., "Use a Fault Tree Approach", Hydrocarbon Processing, 56(9), 1977.

Lambert, H. E. and Gilman, F. M., The IMPORTANCE Computer Code, 11th Loss Prevention Symposium, AIChE, 1977.

Lambert, H. E. and Yadigaroglu, G., Fault Tree Diagnosis of System Fault Conditions, UCRL-78170 (Livermore, CA), 1976.

Lambert, H. E., Fault Trees for Decision Making in Systems Analysis, UCRL-51829 (Livermore, CA), 1975.

Lambert, H. E., Fault Trees for Location of Sensors in Chemical Processing Systems, UCRL-78442 (Livermore, CA), 1976.

Lambert, H. E., "Fault Trees for Locating Sensors in Process Systems", Chemical Engineering Progress, 73(8), 1977.

Lapp, S. A. and Powers, G. J., "Computer-Assisted Generation and Analysis of Fault Trees", Loss Prevention and Safety Promotion, 1977.

Lee, Y. T. and Apostolakis, G. E., Probability Intervals for the Top Event Unavailability of Fault Trees, UCLA-ENG-7663, 1976.

Martin-Solis, G. A., Andow, P. K., and Lees, F. P., "An Approach to Fault Tree Synthesis for Process Plants", Loss Prevention and Safety Promotion, 1977.

Mearns, A. B., Fault Tree Analysis: The Study of Unlikely Events in Complex Systems, System Safety Symposium, 1965.

Mingle, J. O., Chawla, T. O., and Person, L. W., Probabilistic Safety Analysis Methods with Application to Nuclear and Chemical Systems, 11th Loss Prevention Symposium, AIChE, 1977.

Powers, G. J. and Lapp, S. A., "Computer-Aided Fault Tree Synthesis", Chemical Engineering Progress, 72(4), 1976.

Powers, G. J. and Tompkins, F. C., "A Synthesis Strategy for Fault Trees in Chemical Processing Systems", Loss Prevention, 1974.

Powers, G. J. and Tompkins, F. C., "Fault Tree Systems for Chemical Processes", AIChE Journal, 20, 1974.

Vesely, W. E., et al., Fault Tree Handbook, NUREG-0492, 1981.

Event Tree Analysis

Atomic Energy Commission, Reactor Safety Study, WASH-1400, 1975.

Andow, P. K., Event Trees and Fault Tree, Loughborough University of Technology, Course on Loss Prevention in the Process Industries, 1976.

Andow, P. K. and Lees, F. P., "Process Plant Alarm Systems: General Considerations", Loss Prevention and Safety Promotion, 1974.

Rasmussen, J., The Human Data Processor as a System Component: Bits and Pieces of a Model, RISO-M-1722, 1974.

Cause-Consequence Analysis

Himmelblau, D. M., Fault Detection and Diagnosis in Chemical and Petrochemical Processes, Amsterdam: Elsevier, 1978.

Nielsen, D. S., "Use of Cause Consequence Charts in Practical Systems Analysis", in Reliability and Fault Tree Analysis, SIAM, 1975.

Nielsen, D. S., The Cause Consequence Diagram Method as a Basis for Quantitative Accident Analysis, RISO-M-1374, 1971.

Nielsen, D. S., Use of Cause Consequence Charts in Practical Systems Analysis, RISO-M-1743, 1974.

Nielsen, D. S., Platz, O., and Runge, B., "A Cause Consequence Chart of a Redundant Protection System", IEEE Transactions on Reliability, R-24, 1975.

Taylor, J. R., A Semiautomatic Method for Qualitative Failure Mode Analysis, RISO-M-1707, 1974.

Taylor, J. R., Sequential Effects in Failure Mode Anaysis, RISO-M-1740, 1974.

Taylor, J. R., Common Mode and Coupled Failure, RISO-M-1826, 1976.

Taylor, J. R., Cause Consequence Diagrams, NATO Advanced Study Institute, 1978.

Taylor, J. R. and Hollo, E., Algorithms and Programs for Consequence Diagram and Fault Tree Construction, RISO-M-1907, 1977.

Human Error Analysis

Bell, B. J. and Swain, A. D., "A Procedure for Conducting Human Reliability Analysis", NUREG/CR-2254, U.S. Nuclear Regulatory Commission, Washington, D.C., May, 1983.

Edwards, E. and Lees, F. P., The Human Operator in Process Control, London: Taylor and Francis, 1974.

Embrey, D. E., "Approaches to the Evaluation and Reduction of Human Error in the Process Industries", Institution of Chemical Engineers Symposium, 66, 1981.

Huchingson, R. D., New Horizons for Human Factors in Design, McGraw-Hill Book Company, New York, N.Y., 1981.

Kletz, T. A. and Whitaker, G. D., Human Error and Plant Operation, EDN 4099, Safety and Loss Prevention Group, Petrochemicals Division, Imperial Chemical Industries, Ltd., Billingham, England, 1973.

Meister, D., Human Factors: Theory and Practice, John Wiley and Sons, New York, N.Y., 1971.

Rasmussen, J. and Taylor, J. R., Notes on Human Factors Problems in Process Plant Reliability and Safety Prediction, Risø-M-1894, Risø National Laboratory, Roskilde, Denmark, September, 1976.

Rasmussen, J., Human Errors. A Taxonomy for Describing Human Malfunction in Industrial Installations, Risø-M-2304, Risø National Laboratory, Roskilde, Denmark, August, 1981.

Rook, L. W., Reduction of Human Error in Industrial Production, SCTM, 93-62(14), Sandia National Laboratories, Albuquerque, N.M., June, 1962.

Swain, A. D., THERP, Report SC.R.64.1338, Sandia Laboratories, 1964.

Swain, A. D. and Guttmann, H. E., "Handbook of Human Reliability Analysis with Emphasis on Nuclear Power Plant Applications", NUREG/CR-1278, U.S. Nuclear Regulatory Commission, Washington, D.C., August, 1983.

Van Cott, H. P. and Kinkade, R. G. (eds.), Human Engineering Guide to Equipment Design, U.S. Government Printing Office, Washington, D.C., 1972.

Wickens, C. D., Engineering Psychology and Human Performance, Charles E. Merrill Publishing Company, Columbus, Ohio, 1984.

Woodson, W. E., Human Factors Design Handbook, McGraw-Hill Book Company, New York, N.Y., 1981.

Events and Failure Rate Data

Green, A. E. and Bourne, J. R., Reliability Technology, John Wiley and Sons, Inc., 1972.

Lees, F. P., Loss Prevention in the Process Industries, Vol. 2, London: Butterworths, 1980.